PRACTICAL PROBLEMS in MATHEMATICS
for ELECTRICIANS

PRACTICAL PROBLEMS in MATHEMATICS for ELECTRICIANS

Fourth Edition

CRAWFORD G. GARRARD
STEPHEN L. HERMAN

DELMAR PUBLISHERS inc. ®

NOTICE TO THE READER

Publisher does not warrant or guarantee any of the products described herein or perform any independent analysis in connection with any of the product information contained herein. Publisher does not assume, and expressly disclaims, any obligation to obtain and include information other than that provided to it by the manufacturer.

The reader is expressly warned to consider and adopt all safety precautions that might be indicated by the activities described herein and to avoid all potential hazards. By following the instructions contained herein, the reader willingly assumes all risks in connection with such instructions.

The publisher makes no representations or warranties of any kind, including but not limited to, the warranties of fitness for particular purpose or merchantability, nor are any such representations implied with respect to the material set forth herein, and the publisher takes no responsibility with respect to such material. The publisher shall not be liable for any special, consequential or exemplary damages resulting, in whole or in part, from the readers' use of, or reliance upon, this material.

Delmar Staff

Executive Editor: Mark W. Huth
Associate Editor: Jonathan Plant
Managing Editor: Barbara A. Christie
Production Editor: Eleanor Isenhart
Design Coordinator: Susan Mathews

For information, address Delmar Publishers Inc.,
2 Computer Drive West, Box 15-015,
Albany, New York 12212-5015

COPYRIGHT © 1987
BY DELMAR PUBLISHERS INC.

All rights reserved. Certain portions of this work copyright © 1980 and 1973. No part of this work covered by the copyright hereon may be reproduced or used in any form or by any means — graphic, electronic, or mechanical, including photocopying, recording, taping, or information storage and retrieval systems — without written permission of the publisher.

Printed in the United States of America
Published simultaneously in Canada
by Nelson Canada,
A Division of International Thomson Limited

10 9 8 7 6 5 4 3

Library of Congress Cataloging-in-Publication Data

Garrard, Crawford G.
  Practical problems in mathematics for electricians.

  1. Electric engineering—Mathematics. I. Herman, Stephen L.
II. Title.
TK153.G35   1987   513'.13'0246213   86-23926
ISBN 0-8273-2553-3 (pbk.)
ISBN 0-8273-2554-1 (Instructor's guide)

CONTENTS

PREFACE / vii
To the Student — Safety Practices for Electricians / viii

SECTION 1 WHOLE NUMBERS / 1

Unit 1   Addition of Whole Numbers / 1
Unit 2   Subtraction of Whole Numbers / 5
Unit 3   Multiplication of Whole Numbers / 8
Unit 4   Division of Whole Numbers / 11
Unit 5   Combined Operations with Whole Numbers / 14

SECTION 2 COMMON FRACTIONS / 17

Unit 6   Addition of Common Fractions / 17
Unit 7   Subtraction of Common Fractions / 22
Unit 8   Multiplication of Common Fractions / 25
Unit 9   Division of Common Fractions / 28
Unit 10 Combined Operations with Common Fractions / 30

SECTION 3 DECIMAL FRACTIONS / 33

Unit 11 Addition of Decimal Fractions / 33
Unit 12 Subtraction of Decimal Fractions / 37
Unit 13 Multiplication of Decimal Fractions / 40
Unit 14 Division of Decimal Fractions / 43
Unit 15 Decimal and Common Fraction Equivalents / 46
Unit 16 Combined Operations with Decimal Fractions / 49

SECTION 4 PERCENTS, AVERAGES, AND ESTIMATES / 52

Unit 17 Percent and Percentage / 52
Unit 18 Interest / 55
Unit 19 Discount / 58
Unit 20 Averages and Estimates / 60
Unit 21 Combined Problems on Percents, Averages, and Estimates / 63

vi CONTENTS

SECTION 5 POWERS AND ROOTS / 65

Unit 22 Powers / 65
Unit 23 Roots / 68
Unit 24 Combined Operations with Powers and Roots / 73

SECTION 6 MEASURE / 75

Unit 25 Length Measure / 75
Unit 26 Area Measure / 79
Unit 27 Volume and Mass Measure / 84
Unit 28 Energy and Temperature Measure / 89
Unit 29 Combined Problems on Measure / 92

SECTION 7 RATIO AND PROPORTION / 94

Unit 30 Ratio / 94
Unit 31 Proportion / 96
Unit 32 Combined Operations with Ratio and Proportion / 98

SECTION 8 FORMULAS / 99

Unit 33 Representation in Formulas / 99
Unit 34 Rearrangement in Formulas / 102
Unit 35 General Simple Formulas / 104
Unit 36 Ohm's Law Formulas / 113
Unit 37 Power Formulas / 119
Unit 38 Combined Problems on Formulas / 121

SECTION 9 TRIGONOMETRY / 124

Unit 39 Pythagorean Theorem / 124
Unit 40 Trigonometric Functions / 126
Unit 41 Plane Vectors / 129
Unit 42 Rotating Vectors / 135
Unit 43 Combined Problems on Trigonometry / 139

ACHIEVEMENT REVIEW A / 141
ACHIEVEMENT REVIEW B / 145
DIAGNOSTIC READING SURVEY / 150
APPENDIX / 165
GLOSSARY / 179
ANSWERS TO ODD-NUMBERED PROBLEMS / 181

PREFACE

Practical Problems in Mathematics for Electricians is one in a series of widely-used books designed to offer students, trainees, and others practical problem-solving experience in a particular trade area. This new edition has been improved by adding concise mathematical explanations, including worked-out example problems, at the beginning of each unit. This new feature provides the reader a chance to review the procedures needed to solve problems, and greatly increase the flexibility of the book.

The book is an excellent supplement to any vocational mathematics textbook in which more detailed treatment of appropriate mathematics topics can be found. Electrical students will find excellent opportunities to test and develop their problem-solving abilities, and will also receive a valuable review of electrical terminology in the process.

Practical Problems in Mathematics for Electricians offers many benefits for both instructors and students. Two achievement reviews are included at the end of the book to provide an effective means of measuring student progress. A diagnostic reading survey is included, which can be used to determine a student's level of basic skill. A glossary is included to aid students with technical and mathematical definitions, and the appendix provides information on measurement, formulas, and trigonometric functions. Answers to odd-numbered problems are also provided.

An instructor's guide provides the answers to all problems, along with other instructional aids that will be helpful to the teacher.

The late Crawford G. Garrard taught at the Augusta Area Technical School in Georgia, and was active in the field of technical education. This new edition is dedicated to him.

Stephen L. Herman currently teaches in the industrial electricity program at Lee College in Baytown, Texas. He has extensive experience as a teacher and industrial electrician, and has written several other textbooks for electrical trades students.

The following people provided valuable reviews of this revision: O. G. Albright, Dr. Ann E. Orletsky, and Nancy Raulerson.

TO THE STUDENT

SAFETY PRACTICES FOR ELECTRICIANS

1. Be sure the power in the circuit is disconnected. An electrical circuit is safe only after it has been unplugged and the capacitors have been discharged.
2. Be sure your hands are dry when handling an electrical or electronics device that is connected to a live power line. Perspiration may cause your hands to be damp. Any amount of moisture on the surface of the skin reduces resistance to current, thereby increasing the danger of severe shock.
3. Use only those tools and equipment that are in good condition. Replace defective cords and plugs immediately.
4. Use only those electrical or electronic devices which have been approved by the Underwriters Laboratories, Incorporated. This will ensure the maximum degree of safety under the conditions for which the product was designed to operate.
5. Use only carbon dioxide or dry chemical fire extinguishers for the control of fires involving electrical equipment that is connected to power lines.
6. Work on an energized circuit with only one hand. Keep the other hand behind your back or in a pocket. This procedure will prevent current from passing through the chest region of your body.
7. Use an isolation transformer in conjunction with the operation of AC-DC equipment. This procedure reduces the danger of shock and damage to instruments.
8. Be sure to have another person nearby when working on equipment that may be electrically "hot."
9. Be sure to stand on a dry surface when using electrical or electronic devices. Rubber-soled shoes or a plastic mat will avoid a direct path to the ground.
10. Use caution and common sense when working with electricity.

RESCUE OF SHOCK VICTIMS

In case of an electrical accident, the victim may not be able to free oneself from contact with the associated wires, terminals and electrodes. Since time is of extreme importance, the first step to be taken is to promptly remove the victim from the contact. While doing this, the following precautions should be observed in order to prevent the rescuer from becoming an additional victim:

1. Do not touch the person with your bare hands until you are certain that the associated circuit has been broken or turned off.
2. If the circuit cannot be turned off, use a dry wooden stick or other insulator material to free, or, if necessary, to knock the victim from the contact. If none of these items is readily available, cover your hands thoroughly with dry clothing and, while standing upon a dry insulator material, push the victim away. Be certain that your body does not come into contact with wires or terminals.

Mouth-to-Mouth Resuscitation

After the victim has been removed from electrical contact, check for breathing. If breathing is abnormal, start artificial respiration immediately. Do not waste time. Seconds count.

The most effective procedure of administering artificial respiration is the mouth-to-mouth method, which is as follows:

1. Turn victim on back.
2. If necessary, clear victim's mouth, nose and throat of mucus or foreign objects so that the air passageway to the lungs is open. This can be done with a cloth or fingers.
3. Place victim's head back as far as possible so that the front of the neck is stretched (see drawing A). This position of the head further clears the air passageway by causing the tongue to move away from the back of the throat.
4. Open your mouth and place it firmly over victim's open mouth. At the same time, pinch victim's nostrils shut (see drawing B).

Principal Steps For Mouth-to-Mouth Resuscitation

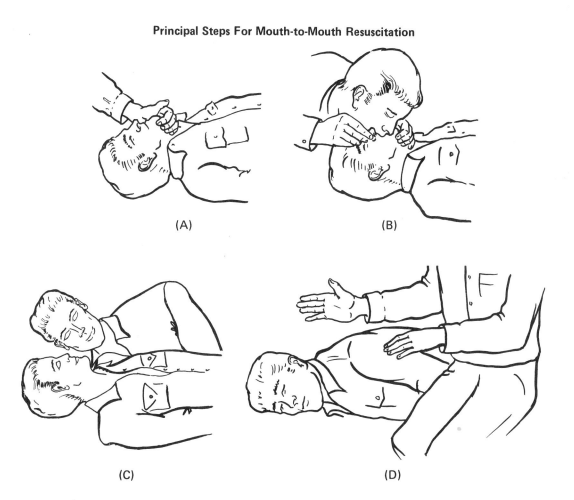

(A)　　　　(B)

(C)　　　　(D)

5. Vigorously blow air into the victim's mouth. Then remove your mouth and turn your head to the side to watch the victim's chest and listen for signs of breathing (see drawing C). If the chest rises or you hear the return rush of air, this indicates that air exchange is taking place. Note: If air exchange does not occur, turn victim on side and, with the palm of your hand, strike several sharp blows between shoulder blades (see drawing D). Turn victim on back once again. Recheck head and neck position.
6. If victim is an adult, vigorously blow air into mouth at the rate of approximately 12 breaths per minute. If victim is a child, blow air less vigorously at the rate of approximately 20 breaths per minute.
7. Continue artificial respiration until victim is breathing normally or until a physician instructs you to stop.
8. After victim has been revived, victim should be kept warm and as quiet as possible. Do not feed or give victim liquids of any kind immediately following the procedure.
9. Call a doctor for a complete evaluation on the condition of victim.

Remember, if one is to apply emergency resuscitation effectively, one must know what to do and must do it immediately.

Whole Numbers

 Unit 1 ADDITION OF WHOLE NUMBERS

BASIC PRINCIPLES OF ADDITION OF WHOLE NUMBERS

Whole numbers refer to complete units with no fractional parts. Addition is the process of finding the *sum* of two or more numbers. Whole numbers are added by placing them in a column with the numbers aligned on the right side of the column. The right column of numbers is added first. Write the last digit of the sum in the answer. The remaining digit(s) is carried to the next column and added. This procedure is followed until all columns have been added.

Example: Find this sum. 25 + 7 + 126 + 54 + 367

```
      2                12               1 2
     25                25                25
      7                 7                 7
    126               126               126
     54                54                54
  + 367             + 367             + 367
  -----             -----             -----
      9                79               579
```

PRACTICAL PROBLEMS

1. When taking inventory, the numbers of BX connectors found in five different bins are 176, 264, 375, 234, and 116. What is the total number of connectors in all bins? _____

2. In eight different boxes there are a number of 3/4-inch, #8 flat head, bright wood screws. The numbers of screws are 124, 72, 36, 92, 38, 64, 74, and 67. What is the total number of screws? _____

3. In wiring eight houses, outlets are installed. The graph shows the number of outlets installed in each house. Find the total number of outlets that must be roughed-in. _____

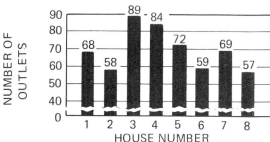

4. An electrician uses switch outlet boxes on eight different jobs. The number of boxes used on each job is 56, 9, 86, 36, 93, 105, 42, and 56. Find the total number of outlet boxes used. _____

5. The materials charged to a wiring job are as follows: 100 ampere distribution panel, $36; meter switch, $8; conduit, $28; number 2 wire, $43; BX cable, $25; conduit fittings, $9; outlet boxes, $92; switches, $35; fixtures, $65; and $37 for wire nuts, grounding clips, staples, and pipe clamps. What is the total amount charged for these materials? _____

6. At different times during a week, an electrician takes the following amounts of metallic cable from stock: 500 feet, 1 200 feet, 250 feet, 90 feet, 38 feet, 65 feet, 84 feet, 225 feet, and 125 feet. What is the total number of feet of metallic cable taken from stock? _____

7. The following amounts of non-metallic cable are used on an apartment house job: 625 metres, 785 metres, 75 metres, 140 metres, 310 metres, 325 metres, and 120 metres. What is the total number of metres of non-metallic cable used on the job? _____

8. A factory department has motors of 75 horsepower, 30 horsepower, 200 horsepower, 40 horsepower, 25 horsepower, 15 horsepower, 5 horsepower, 125 horsepower, 150 horsepower, and 175 horsepower. What is the combined horsepower of the 10 motors? _____

9. An electrical supply house purchases in separate lots 35 pounds, 40 pounds, 125 pounds, 200 pounds, 75 pounds, 90 pounds, 20 pounds, and 30 pounds of solder. Find the total number of pounds of solder purchased. _____

10. The line graph shows the monthly consumption of energy in kilowatt-hours for a house during a one-year period. Find the total amount of energy consumed during the year. _____

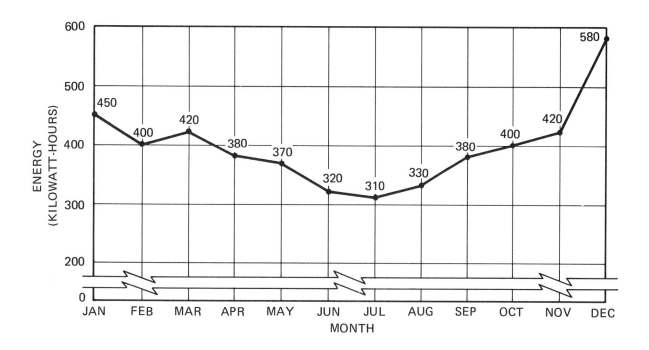

11. A school has twelve lighting circuits which use the following: 545 watts, 650 watts, 750 watts, 1 820 watts, 2 462 watts, 2 571 watts, 1 360 watts, 1 540 watts, 793 watts, 1 225 watts, 330 watts, and 793 watts. What is the total number of watts consumed when all these circuits are being used? _____

12. The cost of magnet wire for a motor repair shop during a one-week period is as follows: 14 pounds of number 17 — $58; 12 pounds of number 16 — $55; 10 pounds of number 24 — $51; 6 pounds of number 21 — $19; 5 pounds of number 25 — $24. Find the total cost of magnet wire during this period. _____

4 Section 1 Whole Numbers

13. From a full container of dry cells, 325 dry cells are placed in the stockroom, 45 dry cells are placed on the shelf in the showroom, and 18, 25, 30, 24, and 6 dry cells are sold to customers. How many dry cells are taken from the full container? _____

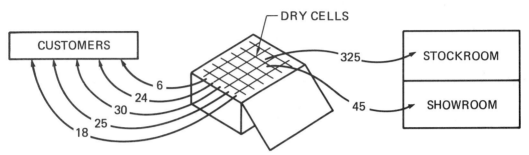

14. Three rooms of a house contain lamps which have the following wattage: living room, 150 watts; dining room, 125 watts; bathroom, 75 watts. What is the total load when all lamps are operating? _____

15. The following number of BX cable staples are used during a given period: 250, 125, 65, 36, 48, 96, 92, 28, 42, 106, 140, and 24. Find the total number of BX cable staples used during this period. _____

16. An electrical contractor receives quantities of Braidx cable during the first quarter of the year: January 1—7 500 feet; February 1—10 750 feet; March 1 — 4 500 feet. Find the total number of feet of Braidx cable received. _____

17. During one week of work an electrician used the following amounts of three conductor with ground NM cable: 1 200 centimetres, 1 150 centimetres, 1 076 centimetres, 180 centimetres, and 100 centimetres. Find the total amount of cable used. _____

Unit 2 SUBTRACTION OF WHOLE NUMBERS

BASIC PRINCIPLES OF SUBTRACTION OF WHOLE NUMBERS

Subtraction is the process of finding the *difference,* or *remainder,* between two numbers. The smaller of the two numbers is placed below the larger, keeping the right column of numbers aligned.

Example: Subtract 432 from 768.

```
  768
- 432
  336
```

If the digit being subtracted is larger than the top digit, borrow 1 from the digit in the column to the left as shown in the following example.

Example: 853 − 38

```
        (5 becomes 4)
  853   (3 becomes 13)
-  38
  815
```

PRACTICAL PROBLEMS

1. An electrician removes from stock 500 feet of BX cable on Monday, 250 feet on Tuesday, and 750 feet on Wednesday. On Friday, 339 feet of BX cable are returned. How many feet of BX cable are used? _____

2. An electrical contractor charges $239 for a job. The materials cost $105. The cost of labor is $69 and the cost of transportation is $5. Find the profit. _____

3. An inventory sheet shows 565 outlet boxes on January 1. On January 10, 145 boxes are taken out of stock. On January 14, 35 boxes are returned to stock. How many outlet boxes are in stock after January 14? _____

6 Section 1 Whole Numbers

4. For a residential job, a reel containing 1 050 feet of cable is delivered. Three 45-foot lengths and three 65-foot lengths are used. How many feet are left? _____

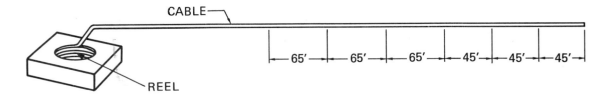

5. A coil of Type S cord, 90 metres long, is taken on a job. The lengths cut off for drop and extension cords are 30 metres, 6 metres, 3 metres, 25 metres, 15 metres, and 2 metres. How many metres of cord remain in the coil? _____

6. A 1 000-foot reel of large stranded cable weighs 1 106 pounds. Of this, 365 pounds are used on a certain job from the switch to the first pull box, and 422 pounds are used from the first box to the last box. How many pounds of wire are left on the reel? _____

7. A purchase of 2 500 feet of number 14 wire is made for a job. On November 1, 1 365 feet of this wire are used. On November 3, an additional 830 feet are used. How many feet of wire are left after November 3? _____

8. On a certain job, a sum of $438 is spent for materials. Of this amount, $76 is spent for 1-inch conduit, and $105 is spent for cable. How much money is spent for other materials? _____

9. During the month of November, 400 outlet boxes are purchased at a cost of $152. The numbers of outlet boxes used are as follows: on December 1 — 59 boxes; on December 5 — 69 boxes; and on December 12 — 72 boxes. How many outlet boxes are left? _____

10. At inventory time, 435 pounds of magnet wire for winding motors are checked as being available. In ten successive days, 15 pounds, 6 pounds, 24 pounds, 12 pounds, 3 pounds, 8 pounds, 17 pounds, 32 pounds, 16 pounds, and 13 pounds of wire are taken out of stock. How many pounds are left? _____

11. A customer receives an electricity bill. The bill states that 1 876 kilowatt-hours of energy are used. Of this total, 504 kilowatt-hours are used for lighting and the rest are used for hot water. How many kilowatt-hours does the customer use for hot water? _____

Unit 2 Subtraction of Whole Numbers 7

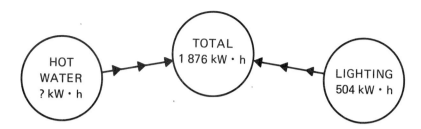

12. A supply house has 804 solenoids for 5-horsepower motor controls. The clerk must reorder this item when the supply reaches 60. On January 1, 14 are sold, and on January 16, 75 are sold. How many more can be sold before reordering? _____

13. A bin contains a total of 173 octagon boxes. For two jobs, 47 boxes and 65 boxes are taken from the bin. One job uses 4 boxes less than originally estimated, and these 4 are returned to the bin. How many boxes are now in the bin at the end of the two jobs? _____

14. A buyer can purchase 70 screwdrivers. Ten 4-inch lengths, twelve 6-inch lengths, twenty 8-inch lengths, and twenty 10-inch lengths are needed. How many heavy 24-inch length screwdrivers can be bought and obtain the total of 70 screwdrivers? _____

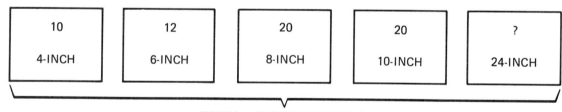

15. An electrician is given a 250-foot coil of BX cable. On one run 29 feet are used. How much BX cable is returned to stock? _____

16. A two-wire transmission line requires 134 miles of conductor for each wire. By straightening the proposed right of way, the distance is reduced by 7 miles. What will be the length of each new wire? _____

17. A total resistance of 60 megohms is needed. On hand are three resistors with the following values: 14 megohms, 25 megohms, and 11 megohms. What is the value of the additional resistor required? _____

18. A tapered pin has a small-end diameter of 101 centimetres and a large-end diameter of 189 centimetres. What is the difference between the two diameters? _____

Unit 3 MULTIPLICATION OF WHOLE NUMBERS

BASIC PRINCIPLES OF MULTIPLICATION OF WHOLE NUMBERS

Multiplication is actually a simple method of addition. For example, if four 5s are added, the answer will be 20. If the number 5 is multiplied by 4, the answer (known as the *product*) is equal to 20. Therefore, 5 × 4 is the same as adding four 5s.

$$
\begin{array}{r} 5 \\ 5 \\ 5 \\ +\ 5 \\ \hline 20 \end{array} \qquad \begin{array}{r} 5 \\ \times\ 4 \\ \hline 20 \end{array}
$$

To multiply more complex numbers, write the number to be multiplied. Then write under it the number of times it is to be multiplied. In the following example, the number 247 is to be multiplied by 32. Write the numbers keeping the right column aligned.

Example: 247 × 32

```
                                        2       12      12
   1        1        1        1         1        1       1
  247      247      247      247       247      247     247
 X 32     X 32     X 32     X 32      X 32     X 32    X 32
 ----     ----     ----     ----      ----     ----    ----
    4       94      494      494       494      494     494
                                         1       41     741
                                                       7904
```

Multiply the first two numbers (2 × 7 = 14). Place the 4 below the 2, and carry the one to the next column. Then multiply (2 × 4 = 8) and add the 1 (8 + 1 = 9). Place the 9 beside the 4. Multiply (2 × 2 = 4). Place the 4 beside the 9. Next, multiply each number by the 3 in 32. The answers will be brought down in the same manner except that one space is skipped. Multiply (3 × 7 = 21). Place the 1 below the 9 and carry the 2 to the next column. Multiply (3 × 4 = 12). Add the 2 from the first column (12 + 2 = 14). Place the 4 beside the 1 and carry the 1 to the next column. Multiply (3 × 2 = 6) and add 1 (6 + 1 = 7). Place the 7 beside the 4. The final step is to add the two sets of products together to obtain the total.

PRACTICAL PROBLEMS

1. A panel board requires sixteen 1/2-inch holes, twenty-one 1/4-inch holes, and eleven 5/16-inch holes. Each hole requires a bolt with three washers and two nuts.
    a. Find the total number of washers needed for the 1/2-inch holes.   a. _____
    b. Find the total number of washers needed for the 1/4-inch holes.   b. _____

Unit 3 Multiplication of Whole Numbers 9

    c. Find the total number of washers needed for the 5/16-inch holes.   c. _____
    d. Find the total number of nuts needed for the 1/2-inch holes.   d. _____
    e. Find the total number of nuts needed for the 1/4-inch holes.   e. _____
    f. Find the total number of nuts needed for the 5/16-inch holes.   f. _____

2. A bearing on a large machine is tested over a period of 8 hours, at a speed of 40 500 revolutions per hour. How many revolutions does the shaft turn in the bearing during the test period? _____

3. Find the total amount of power, in watts, for the three motors shown, if 1 horsepower equals 746 watts. _____

 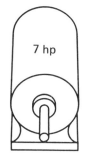 

5 hp   7 hp   10 hp

4. A very small magnet is wound with 97 layers with 215 turns per layer. How many turns of wire are on the coil? _____

5. A coil requires 2 900 turns of number 14 wire. If each of 20 layers is wound with 143 turns, will this satisfy the requirements for the coil? _____

6. A building uses the following size lamps: sixteen 50-watt, nine 15-watt, twelve 25-watt, six 75-watt, and four 100-watt. How many watts are consumed when all the lights in the building are burning? _____

7. It is found that a certain electrical circuit having a total load of 2 800 watts in lamps must be reduced. Ten 200-watt lamps are replaced with ten 150-watt lamps; eight 100-watt lamps are replaced with eight 60-watt lamps. What is the total number of watts in connected lamps after the change is made? _____

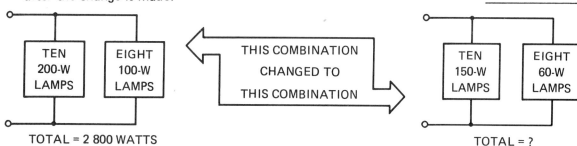

8. Thirty-four steel boxes are used for a certain wiring job. In each box five 1-inch holes are drilled. In twenty-three of the boxes, two 1 3/4-inch holes are drilled, and the remaining boxes have three 1 1/2-inch holes drilled in them. How many holes are drilled in all the boxes? _____

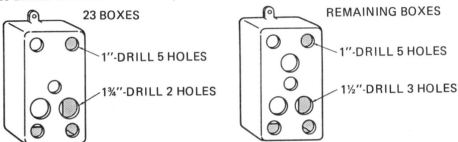

9. An electrical company has a payroll of 27 people. Seven people earn $8 per hour, eleven people earn $10 per hour, and nine people earn $6 per hour. If all employees work 40 hours during the week, what is the total amount of money that is earned in one week? _____

10. The product of current (amperes) and voltage (volts) equals power (watts). The total power of an electrical circuit is equal to the sum of the individual powers. Find the total power for the circuit shown. _____

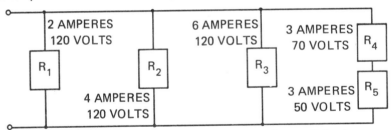

11. A large room contains 40 fluorescent lamps. Twenty-three of the lamps are 40-watt lamps, and the remaining number are 60-watt lamps. What is the total amount of watts used by the lamps? _____

12. A certain lighting job requires four incandescent fixtures, twelve direct fluorescent luminaires, and nine semidirect fluorescent luminaires. The incandescent fixtures cost $19 each, the direct luminaires cost $26 each, and the semidirect luminaires cost $31 each. Find the total cost of all the fixtures and luminaires. _____

13. A nine-floor apartment building has an average of four electrical circuits for each apartment. There are six apartments on each floor, and one on the roof. Find the total number of electrical circuits in the apartment building. _____

14. An electrician uses 763 metres of conduit on each floor of a five-story building. What is the total length of conduit used? _____

Unit 4   DIVISION OF WHOLE NUMBERS

BASIC PRINCIPLES OF DIVISION OF WHOLE NUMBERS

In unit 3, it was shown that multiplication is actually the process of adding a number together many times. Division is just the opposite. Division is actually the process of subtracting a smaller number from a larger number many times. The number to be divided is referred to as the *dividend*. The number used to indicate the number of times the dividend is to be divided is called the *divisor,* and the answer is known as the *quotient*.

To begin the process of division, the dividend is placed inside the division bracket, the divisor is placed to the left of the dividend, and the quotient is placed above the dividend.

$$\text{Divisor} \overline{)\text{Dividend}}^{\text{Quotient}}$$

Example: Divide 1140 by 17.

```
      6            6           67          67 R1
17)1140      17)1140     17)1140     17)1140
   102          102         102         102
    12          120         120         120
                            119         119
                                          1
```

Place the number 1140 under the division bracket and the number 17 to the left of it. The number 17 cannot be divided into a number which is smaller than itself. Therefore, 17 is divided into the number 114 first. Find what number multiplied by 17 will come the closest to 114 without going over 114. In this example, 6 is that number (6 × 17 = 102). The number 102 is placed below 114 and the two numbers are subtracted from each other. This leaves 12. 17 cannot be divided into 12, so the next number of the dividend is brought down to the right of 12. When the zero is placed beside 12, 17 is then divided into 120. The nearest number that 17 can be multiplied by and not go over 120 is 7 (7 × 17 = 119). 119 is placed below 120 and the two numbers are subtracted from each other. This leaves a remainder of 1. Since there are no more numbers in the dividend, the 1 is taken to the quotient and shown as R1 which means a "remainder" of 1.

12 Section 1 Whole Numbers

PRACTICAL PROBLEMS

1. In a 184-foot run of BX cable, the staples are placed 4 feet apart. How many staples are used if one staple is placed at the beginning and one is placed at the end of the run? _____

2. An electrical contractor purchases 15 fittings of one type for $45 and 6 of another type for $36.
   a. Find the cost per fitting for those costing $45.   a. _____
   b. Find the cost per fitting for those costing $36.   b. _____

3. A total load of 25 620 watts is distributed equally over the branch circuits shown. What is the average load per circuit in watts? _____

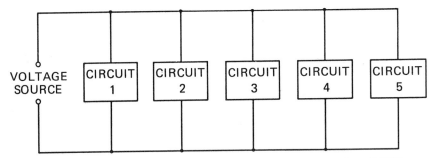

4. How many 250-foot rolls of BX cable are needed if a job requires 5 250 feet? _____

5. A hotel with 22 rooms on each of the seven floors has a total of 770 outlets. If each room has the same number of outlets, how many are there in each room? _____

6. A certain wiring job has 28 outlets equally spaced for 351 feet. If one outlet is placed at the beginning and one at the end, what is the center-to-center distance between outlets? _____

7. In a house where 35 outlets are installed, 735 feet of cable are used. What is the average number of feet of cable used per outlet? _____

8. Twelve standard packages of conduit fittings are purchased. What is the weight per package? _____

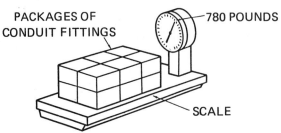

9. Two electricians work a total of 640 hours on a job. Each works 8 hours per day, 5 days per week. How many weeks does each electrician work?

10. Two thousand, five hundred feet of BX cable are ordered. The cable is shipped in 250-foot coils. How many coils are shipped?

11. A certain machine room uses 2 160 watts to supply 60-watt lamps for bench lighting. How many lamps are connected?

12. A wiring job uses 1 232 metres of cable for 56 outlets. What is the average number of feet per outlet?

13. The cost for the carton of staples shown is $36. What is the cost per standard package?

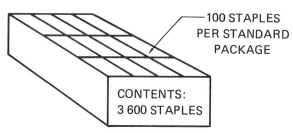

14. On a 124-foot length of Romex cable, 32 staples are used. The staples are equally spaced. If one staple is placed at the beginning and one at the end of the cable, how far apart are the staples placed?

15. A total load of 15 840 watts is distributed equally over 12 circuits. What is the load per curcuit in watts?

16. Box A and box B each contain type C connectors. Box A contains 200 connectors and costs $30. Find the cost of box B which contains 250 connectors. The unit price is the same for both boxes.

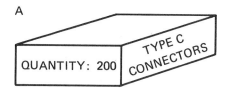 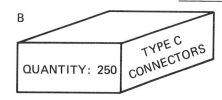

17. A school study hall is 90 metres by 90 metres. Two rows of lighting are placed in the ceiling. What is the center-to-center distance between the rows of lights if the rows are spaced with equal distances from the side walls and between the rows?

Unit 5 COMBINED OPERATIONS WITH WHOLE NUMBERS

BASIC PRINCIPLES OF COMBINED OPERATIONS

This unit provides practical problems involving combined operations of addition, subtraction, multiplication and division of whole numbers.

PRACTICAL PROBLEMS

1. In wiring eight houses, the electricians install 68, 87, 57, 74, 49, 101, 99, and 56 outlets. Find the total number of outlets that must be roughed-in. _____

2. An electrician removes from stock, at different times, the following amounts of BX cable: 120', 327', 637', 302', 500', 250', 140', 75', and 789'. Find the total number of feet of BX cable taken from stock. _____

3. An electrical supply house purchases in separate lots, 30, 120, 37, 125, 103, 33, 210, and 40 pounds of solder. What is the total number of pounds of solder purchased? _____

4. A school has twelve electrical circuits which have a capacity of 2 569, 1 260, 1 639, 563, 790, 800, 1 137, 250, 500, 750, 1 830, and 2 462 watts. What is the total number of watts consumed when all these circuits are being used under their total loads? _____

5. BX cable in the following amounts is used on an apartment house job: 250 feet, 71 feet, 39 feet, 110 feet, 75 feet, 87 feet, and 560 feet. What is the total amount of cable used on the job? _____

6. The following number of BX staples are used during a given period: 28, 250, 38, 108, 92, 130, 25, 36, 97, 91, 65, and 40. Find the total number of BX staples used. _____

7. An electrician takes out of stock 498 feet of BX cable on Monday, 103 feet on Tuesday, and 78 feet on Wednesday. On Friday, 27 feet of BX cable are returned to stock. How much BX cable is used? _____

8. An inventory sheet shows a balance of 500 outlet boxes on January 1. On January 10, 127 outlet boxes are taken out of stock. On January 14, 61 outlet boxes are returned to stock. How many outlet boxes are left in stock? _____

DESCRIPTION:				TITLE: OUTLET BOXES			
1. Joe's Electrical Supplies 2. 3.				MAX.		MIN.	
DATE	IN	OUT	BAL.	DATE	IN	OUT	BAL.
1/1			500				
1/10		127	?				
1/14	61		?				

9. An electrical contractor charges $350 for a job. The materials cost $105. The cost of labor is $139 and the cost of transportation is $11. Find the profit.

10. A purchase of 2 500 feet of number 14, double-braided, rubber-covered wire is made for a job. On November 1, 978 feet of this wire are used, and on November 3, 1 023 feet are used. How many feet of wire are left?

11. A building contains seventy 100-watt lamps, thirty-eight 75-watt lamps, ten 60-watt lamps, and twenty 40-watt lamps. If all lamps are on at the same time, how many watts are used?

12. An electrical contractor employs 16 people. Five people earn $5 per hour, four people earn $7 per hour, and the remaining people earn $6 per hour. What is the total hourly wage earned by all 16 people?

13. A 7-floor apartment building has an average of 7 electrical circuits per apartment, and there are 8 apartments per floor. How many electrical circuits are there in the building?

14. A wiring job requires 5 127 feet of cable. If the cable comes in 250-foot coils, how many coils of cable are required?

15. A coil of wire is wound in 7 layers with 13 turns per layer. How many turns of wire are on the coil?

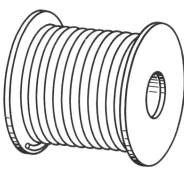

16. A wiring job requires 29 outlets are to be spaced equally over 364 feet. One outlet is placed at the beginning of the 364 feet and one at the end. Find the center-to-center distance between the outlets. _____

17. An order is placed for 16 coils of cable. The cable comes in 250-foot coils. How many feet of cable are received? _____

18. Twenty standard cartons of octal boxes weigh a total of 1 100 pounds. Find the weight per carton. _____

Common Fractions

SECTION 2

Unit 6 ADDITION OF COMMON FRACTIONS

BASIC PRINCIPLES OF ADDITION OF COMMON FRACTIONS

Measurement must often be closer than can be done with whole numbers. One method of indicating quantities which are smaller than a whole is with common fractions. There are two parts to a common fraction, the *numerator* (the number above the line) and the *denominator* (the number below the line.)

The denominator indicates the number of equal parts the whole is divided into. The inch, for example, is often divided into 16 equal parts. The numerator indicates the number of parts used. If a measurement is 5/16 inch, it indicates that an inch has been divided into 16 equal parts and the length corresponds to 5 of these 16 parts.

Adding Fractions

When fractions are added, their denominators must all be the same. When all the denominators are the same, only the numerators are added.

Example: 1/16 + 3/16 + 5/16

$$\begin{array}{r} \frac{1}{16} \\ \frac{3}{16} \\ + \frac{5}{16} \\ \hline \frac{9}{16} \end{array}$$

If all the denominators are not the same, it is necessary to find a *common denominator*. A common denominator can be found by finding some number that all the denominators of the individual fractions will divide into. In the following example, a common denominator could be 24 since the denominator of each fraction will divide into 24 an equal number of times. This is not to say that 24 is the only common denominator. 48 could also be used as a common denominator. 24, however is the lowest common denominator (LCD). It is generally preferred to use the LCD when adding fractions.

Example: 5/12 + 1/4 + 1/6 + 3/24

$$\frac{5}{12} = \frac{10}{24}$$
$$\frac{1}{4} = \frac{6}{24}$$
$$\frac{1}{6} = \frac{4}{24}$$
$$+ \frac{3}{24} = \frac{3}{24}$$
$$\frac{23}{24}$$

Addition of these fractions is accomplished by changing each fraction into an equivalent fraction with a denominator of 24. This is done by dividing the denominator of the fraction to be changed into the common denominator, and then multiplying the numerator by the answer. 5/12 is changed into 10/24: (24 ÷ 12 = 2); (2 X 5 = 10). The fraction 5/12 and 10/24 are equal in value. To change 1/4 into an equivalent fraction in 24ths: (24 ÷ 4 = 6); (6 X 1 = 6). The fraction 1/4 has the same value as 6/24. When each fraction has been changed into an equivalent fraction in 24ths, the numerators are then added together.

Reducing Fractions to Lowest Terms

It is common practice to reduce a fraction to its lowest terms. This is done by dividing both the numerator and denominator by the same number.

Example: Express 18/48 in lowest terms.

$$\frac{18 \div 6}{48 \div 6} = \frac{3}{8}$$

The fractions 3/8 and 18/48 are equal in value. Some fractions such as 23/24 cannot be reduced because there is no number except 1 that will divide into both 23 and 24 an equal number of times. The fraction 23/24 is in lowest terms.

PRACTICAL PROBLEMS

1. Find the sum of each of the following.
   a. 5/16, 3/8, 3/4, 7/32
   b. 1/16, 3/4, 7/8, 31/32
   c. 3/4, 8/12, 5/8, 5/16
   d. 7/10, 4/5, 3/4, 3/10

   a. _____
   b. _____
   c. _____
   d. _____

Note: Use this diagram for problems 2 and 3.

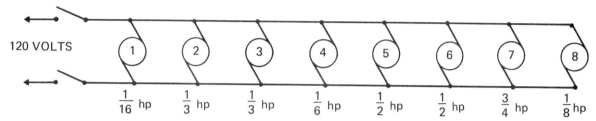

2. Eight motors, each having a rated horsepower as indicated on the diagram, are connected in a circuit. What is the total horsepower of motors 1 to 4 inclusive?

3. What is the total horsepower of motors 5 to 8 inclusive?

4. Eight resistances are connected in series as shown. The unit of measure for resistance is the ohm. The resistance in a series circuit is the sum of all resistances in that circuit. Find the total amount of resistance in the circuit.

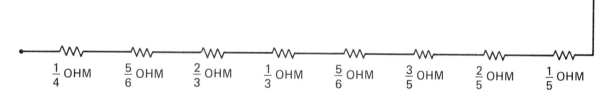

5. The insulation to be used in a certain slot is fish paper 1/64 inch, Tufflex 1/32 inch, varnished cambric 1/64 inch, and top stick 1/8 inch. What is the total thickness of all the insulation?

6. A piece of carbon 31/32 inch thick has a copper plating of 1/64 inch on each side. What is the total thickness?

7. When aligning a motor, pieces of steel are used under the base. The pieces are 13/16 inch thick, 5/64 inch thick, 3/32 inch thick, and 1/8 inch thick. What is the total thickness of all these pieces?

Section 2 Common Fractions

8. A copper contact is insulated from its support with one piece of mica 1/64 inch thick, two pieces of fiber each 1/8 inch thick, and 1/16 inch of pressboard. The copper is 3/4 inch thick and the support is 7/8 inch thick. What is the total thickness of copper, insulation, and support? _____

9. A form is made to wind a coil. The allowances made for insulation are as follows: four layers of 1/64-inch tape, two pieces of 1/16-inch fish paper, and two pieces of 1/8-inch fiber. What is the total thickness allowed for insulation? _____

10. On a theater repair job, it is found that a fixture has been hung from several pieces of wood spiked together. The bracket shown is made to take the place of the wood. The pieces of the bracket are all 5/8 inch thick. What is height H of the bracket? _____

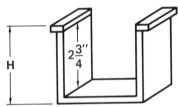

11. What is the total thickness of a wall having 7/8-inch finish siding, 7/8-inch rough siding, 3 3/4-inch studs, and 13/16 inch of lath and plaster? _____

12. Find the distance O for the bend of electrical conduit that is passing under the girder shown in the figure. _____

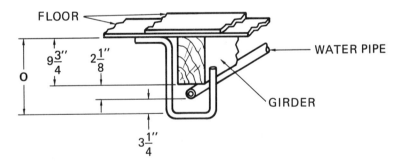

13. What space on a bolt will six washers or spacers of the following thicknesses occupy: 1/16 inch, 1/2 inch, 1/8 inch, 3/32 inch, 3/64 inch, 3/16 inch? _____

14. Find the shortest length of 1/2-inch conduit from which the following pieces can be cut: 7 7/8 inches, 3 1/4 inches, 6 1/2 inches, 12 1/8 inches, and 24 3/16 inches. Allow 1/16 inch for each saw cut. _____

15. What is the shortest strip of fiber from which five pieces of the following lengths can be cut: 7/8 inch, 3 1/8 inches, 2 1/16 inches, 12 1/4 inches, and 7/16 inch? Allow 1/8 inch for each saw cut.

16. Find the total resistance (R_t) of a three series connected resistors.

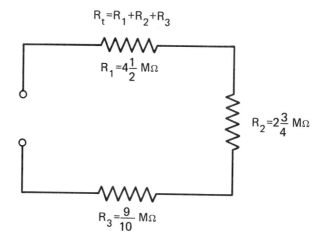

17. Two-wire NM cable is used to connect four outlets. The distances between the outlets are 6 2/10 metres, 8 9/10 metres, 20 1/2 metres, and 8 3/10 metres. Find the total length of wire between outlets.

 # Unit 7 SUBTRACTION OF COMMON FRACTIONS

BASIC PRINCIPLES OF SUBTRACTION OF COMMON FRACTIONS

Subtraction of common fractions is similar to the addition of fractions. In both operations, it is first necessary to have common denominators. Once both fractions have a common denominator, it is then necessary to subtract the numerator of the smaller fraction from the numerator of the larger.

Example: 5/8 − 7/16

$$\frac{5}{8} = \frac{10}{16}$$
$$-\frac{7}{16} = \frac{7}{16}$$
$$\frac{3}{16}$$

In this example, 7/16 is subtracted from 5/8. The lowest common denominator for these two fractions is 16. The fraction 5/8 is changed to 10/16: (16 ÷ 8 = 2); (2 × 5 = 10). The fraction 7/16 is not changed. The numerators of the two fractions are then subtracted from each other (10 − 7 = 3). The answer is 3/16.

PRACTICAL PROBLEMS

1. Two pieces of rigid conduit (electrical pipe) are measured. Each piece has an approximate outside diameter of 1 inch and the thickness of the wall is approximately 1/8 inch. Find the approximate inside diameter of each piece of conduit.

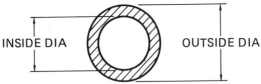

2. The slot shown is 5/16 inch wide and 1/2 inch deep. The piece of fiber is 5/64 inch thick. How deep is the space left (for wires)?

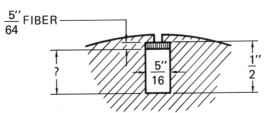

3. A motor with four pieces of steel placed under its base is found to be too high. The total thickness of these pieces is 57/64 inch. It is necessary to remove a piece 3/32 inch thick. What is the total thickness of the steel left under the motor base? _____

4. A washer 1/8 inch thick and a washer 3/32 inch thick are placed on a bolt which is 1 1/2 inches long under the head. What is the distance A left between the two washers after the 1/2-inch nut is screwed on to its full thickness? _____

5. A 1 1/2-inch bolt is inserted through a block 11/16 inch thick with a 1/8-inch washer and a 1/2-inch nut. How much of the bolt will extend beyond the nut when drawn up tightly? _____

6. The length of threading on a BX box connector is approximately 7/16 inch. The locknut is 1/8 inch thick and the metal into which this connector is inserted is 1/32 inch. How much threading is left for a threaded bushing? _____

7. The diameter of a steel shaft, through wear, is reduced 4/1 000 inch. The shaft measured 875/1 000 inch originally. The electrician removes another 8/1 000 inch and makes a new bearing to fit the shaft. What is the new size of the shaft? _____

8. A motor base must be blocked up 6 1/8 inches. Two blocks are used. If one block is 3 3/4 inches thick, how thick is the other? _____

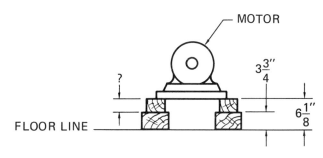

9. A cable measures 500/1 000 inch outside diameter and has 12/1 000 inch of cotton and 50/1 000 inch of rubber insulation. Determine the diameter of the copper conductor (wire).

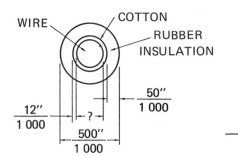

10. A 29/32-inch hole must be enlarged to 59/64 inch to insert a bushing. How much larger than the original hole will this be?

11. A motor commutator 3 1/8 inches in diameter is turned down to remove a flat spot. If 3/64 inch is removed from the surface, find the finished diameter.

12. The motor sleeve bearing shown in the figure has an outside diameter of 3 1/16 inches. The thickness of the wall is 19/64 inch. Find the inside diameter.

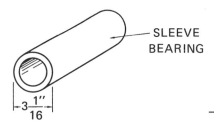

13. A motor brush is 1 7/8 inches long. How long is it after 49/64 inch wears away?

14. Kirchhoff's law is used for circuits when the conductors, forming a part of a network carrying currents, meet at one point. Kirchhoff's law states that the current entering the electrical connection is equal to the sum of the current leaving the connection. In the circuit shown find the current, I_2.

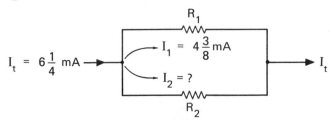

15. An electrician uses 46 3/10 metres of push-back wire from a 200-metre roll. How many metres of wire are left on the roll?

# Unit 8 MULTIPLICATION OF COMMON FRACTIONS

BASIC PRINCIPLES OF MULTIPLICATION OF COMMON FRACTIONS

When fractions are multiplied, it is not necessary to find a common denominator. Fractions can be multiplied by multiplying the numerators together and the denominators together.

Example: Multiply 5/8 by 1/4.

$$\frac{5}{8} \times \frac{1}{4} = \frac{5}{32}$$

When fractions are multiplied, it is often possible to simplify the problem by cross reduction. If a number can be found which will divide an equal number of times into both the numerator of one fraction and the denominator of the other, the problem can be made simpler. The answers should always be reduced to lowest terms.

Example: 4/5 × 15/32

$$\frac{\cancel{4}^{1}}{\cancel{5}_{1}} \times \frac{\cancel{15}^{3}}{\cancel{32}_{8}} = \frac{3}{8}$$

PRACTICAL PROBLEMS

1. The ceiling outlets for the circuit weigh 11/16 pound each. Find the total weight for all the ceiling outlets.

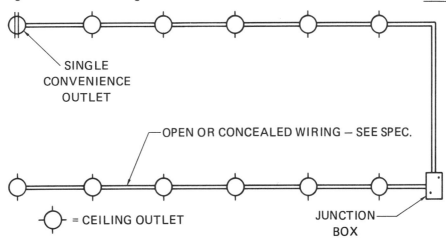

25

26 Section 2 Common Fractions

2. If an electrician works 4 3/4 hours a day on a job at $16.00 per hour, how much money does he earn in 5 days? _____

3. A wiring job calls for thirty-two pieces of 1/2-inch conduit 7 1/2 feet long, eight pieces 13 1/4 inches long, three pieces 7 3/4 inches long, and six pieces 9 5/8 inches long. What is the total number of feet of 1/2-inch conduit required for the job? _____

4. A department in a factory has three 3/4-horsepower motors, five 1/4-horsepower motors, six 3 1/3-horsepower motors, and eight 7 1/2-horsepower motors. What is the total connected motor load in horsepower rating? _____

5. A wiring job requires BX cable in the following lengths: eight pieces 23 1/2 feet long, seven pieces 18 1/2 inches long, twelve pieces 24 1/2 inches long, and twenty-five pieces 19 3/4 inches long. How many feet of BX cable are needed? _____

6. In estimating a job, it is decided that it should take 13 people 3 1/2 hours each to do part of the work, and 7 people 6 3/4 hours each to do the remainder of the job. Determine the total number of hours estimated for the job. _____

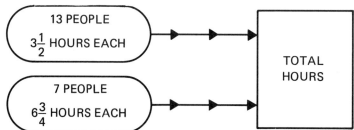

7. For the circuit shown, determine the sum of all the energy used in kilowatt-hours if all units are on for 3 3/4 hours. The power shown for each unit is in kilowatts. _____

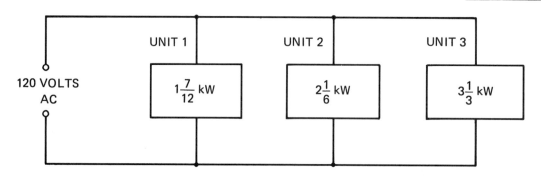

8. A standard package of 3-inch metal molding junction boxes weighs 6 1/4 pounds. If 16 packages are purchased, find the total weight. _____

9. A storage battery contains 1 1/4 kilograms of electrolyte. One-third of the electrolyte is acid. How many kilograms of acid are in the battery cells? _____

10. A 75-watt, 120-volt light uses 2/3 ampere. How many amperes are used by the 5 lights connected in parallel? _____

Unit 9 DIVISION OF COMMON FRACTIONS

BASIC PRINCIPLES OF DIVISION OF COMMON FRACTIONS

Common fractions can be divided in a manner similar to the multiplication of common fractions. When dividing common fractions, it is necessary to invert the divisor and multiply. The same rules that are used for the multiplication of fractions can then be followed. All answers should then be reduced to lowest terms.

Example: Divide 3/4 by 5/8.

$$\frac{3}{4} \div \frac{5}{8} = \frac{3}{\cancel{4}_1} \times \frac{\cancel{8}^2}{5} = \frac{6}{5} = 1\frac{1}{5}$$

PRACTICAL PROBLEMS

Note: No waste allowance is made for cutting unless indicated in the problem.

1. How many 7/8-inch lengths can be cut from a 7-inch fiber strip? _____

2. How many pieces 1 5/16 inches long can be cut from a strip of sheet metal 13 1/8 inches long? _____

3. If 12 6/7 watts are distributed equally over each of the resistors shown, find the average number of watts per resistor. _____

4. A certain circuit is properly fused for a 5-horsepower motor. If the motor is to be replaced by several 5/8-horsepower motors, how many motors will the fuses be able to carry? _____

5. A 20-foot length of underground cable is cut into 6 3/4-inch pieces. How many 6 3/4-inch pieces are made? _____

6. How many whole wedges 3 7/8 inches long can be made from 20 wedge strips, each 3 feet long? _____

7. Two electricians work on an electrical job in an apartment building. Both work five days per week. Over a two-week period, how many times longer does electrician **B** work as compared to electrician **A**? _____

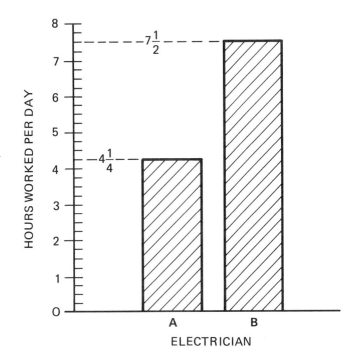

8. How many 1 3/4-inch long machine bolt blanks can be cut from a 5-foot length of stock? Allow 7/32 inch for waste on each blank. _____

9. If 7 3/4 dozen 1/2-inch connector fittings cost $31, what is the cost per dozen? _____

10. One metre of shielded microphone cable weighs 75/1 000 kilogram. What is the length of cable in a coil that weighs 5 1/2 kilograms? _____

Unit 10 COMBINED OPERATIONS WITH COMMON FRACTIONS

BASIC PRINCIPLES OF COMBINED OPERATIONS

This unit provides practical problems involving combined operations of addition, subtraction, multiplication, and division of common fractions.

PRACTICAL PROBLEMS

1. Eight motors are connected in a circuit. The horsepower ratings are 1/8, 3/4, 1/16, 1/2, 1/3, 1/2, 1/3, and 1/4. What is the total horsepower of the circuit? _____

2. The insulation to be used in a certain slot is as follows: fish paper 1/64", Tufflex 1/16", varnished cambric 1/32", and top stick 1/8". What is the total thickness of all the insulation? _____

3. A copper contact is insulated from its support with one piece of mica 1/8" thick, 2 pieces of fiber each 1/16" thick, and 1/8" of pressboard; the copper is 1/32" thick and the support is 1/4" thick. What is the total thickness of copper, insulation, and support? _____

4. When aligning a motor, the following pieces of steel are used under the base: one piece 5/16" thick, one piece 5/8" thick, one piece 5/32" thick, and one piece 5/64" thick. What is the total thickness of all these pieces? _____

5. A form is made to wind a coil. Allowance is made for insulation as follows: 4 layers of 1/32" tape, 2 pieces of 1/8" fish paper, and 2 pieces of 1/4" fiber. What is the total thickness allowed for insulation? _____

6. What space on a bolt will 6 washers, or spacers, of the following thicknesses occupy: 1/64", 1/32", 1/64", 1/32", 1/8", and 3/16"? _____

7. What is the shortest length of 1/2-inch conduit from which the following pieces can be cut: 3 7/8", 5 1/2", 7 3/4", 9 1/8", and 3/8"? Allow 1/64" for saw cuts. _____

8. What is the shortest strip of fiber from which five pieces of the following lengths can be cut: 9 1/2", 10 11/16", 6 9/16", 7 1/2", and 1 5/8"? No allowance is made for waste. _____

9. A slot in a piece of copper is too wide for a wire to be held tightly. The slot is 1/2 inch in width and the wire is 3/8 inch in diameter. How much larger is the slot than the wire? _____

10. The diameter of a steel shaft, through wear, is reduced 6/1 000 inch. The shaft measures 875/1 000 inch originally. The electrician removes another 9/1 000 inch and makes a new bearing to fit the shaft. What is the size of the shaft after the work is completed? _____

11. A cable measures 625/1 000-inch outside diameter and has 16/1 000 inch of cotton and 62/1 000 inch of rubber insulation. What is the diameter of the copper conductor (wire)? _____

12. A 28/32" hole must be enlarged to 58/64" to insert a bushing. How much larger than the original hole will the new hole be? _____

13. A wiring job calls for 36 pieces of 1/2-inch conduit, each 6 1/2 feet long; 9 pieces, each 11 inches long; 4 pieces, each 6 inches long; and 5 pieces, each 18 inches long. What is the total number of feet of 1/2-inch conduit required for the job? _____

14. The ceiling outlets on a residential job weigh an average of 5/8 pounds each and 12 are required. Find the total weight of the outlets. _____

15. A wiring job requires 2/c (2 conductor) BX cable in the following lengths: 5 pieces, each 3 1/2 feet long; 8 pieces, each 18 inches long; and 6 pieces, each 30 inches long. How many feet of 2/c BX are used? _____

16. A department in a factory has three 7 1/2-horsepower motors; four 1 1/2-horsepower motors; two 1/4-horsepower motors; and three 1/2-horsepower motors. What is the total connected motor load in horsepower rating? _____

17. If an electrician works 8 hours a day on a certain job at $10 an hour, what is the pay for a 5-day workweek? _____

18. How many wedges 6 3/4" long can be made from 2 wedge strips, each 5 1/4' long? _____

19. If 7 1/3 yards of varnished cambric (insulation) cost $22, how much does it cost per yard? _____

20. How many 6 3/4-inch lengths can be cut from a 20-foot length of conduit? _____

21. If a motor has a speed of 1 627 3/4 revolutions per second, how many revolutions will it make in 11/15 second? _____

22. Two electricians are assigned to work on a remote-control wiring job. One electrician works 7 1/2 hours each day, and the other electrician works 1 1/2 hours each day. If they both work for 5 days, how many times longer does the first electrician work as compared to the second electrician? _____

23. A box of motor brushes costs $7 1/2. If the price of one brush is $1/4, how many brushes are in the box? _____

24. If 7 1/2 kilowatts of power are distributed equally over 5 resistors, what is the average number of kilowatts per resistor? _____

25. A reel of annunciator wire is purchased for 10 1/2 cents per foot, and the total cost is 525 dollars. How many feet of wire are on the reel? _____

26. A 10 1/2-foot length of electrical metallic tubing is cut into 1 1/2-foot lengths. How many pieces are made? _____

27. An average of 3 1/2 wire nuts is used in each of 14 octal boxes in a home. How many wire nuts are used in all? _____

28. Two electricians work 6 1/2 hours per day for 5 days. Find the total number of hours worked. _____

Decimal Fractions

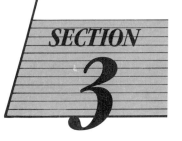

SECTION 3

Unit 11 ADDITION OF DECIMAL FRACTIONS

BASIC PRINCIPLES OF ADDITION OF DECIMAL FRACTIONS

One of the advantages of decimal fractions as compared to common fractions is that when decimal fractions are added or subtracted it is not necessary to find a common denominator. Adding decimal fractions is accomplished by placing the fractions in a column with all the decimal points aligned. The columns are then added in the same manner as adding whole numbers.

Example: 4.0563 + 0.98 + 14.0008 + 0.4005 + 10.33

```
    4.0563
    0.9800
   14.0008
    0.4005
 + 10.3300
   29.7676
```

PRACTICAL PROBLEMS

1. An electrical contractor purchases the following: 750 feet of two-conductor BX cable for $66.50; 250 feet of three-conductor BX cable for $28.38; and 100 feet of 1/2-inch electrical conduit (pipe) for $17.28. What is the total bill for these items? _____

2. What is the total thickness of these shims taken from a bearing: 0.007 inch, 0.125 inch, 0.140 inch, 0.187 inch, and 0.004 inch? _____

3. Materials and labor required to repair an electrical switch are as follows: one shallow box at $0.55; one 3/8-inch hickey at $0.65; one flush toggle switch at $0.98; and labor at $8.92. What is the total cost to repair an electrical outlet? _____

34   Section 3 Decimal Fractions

4. The total current (amperes) is equal to the sum of the individual currents. Find the total current in the figure shown.   _____

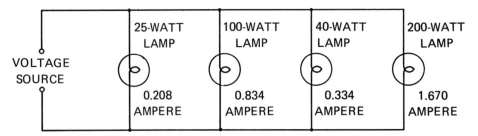

5. What is the cost of winding an alternator if it takes 1 pound of number 16 wire at $3.98; 1 pint of insulating varnish at $1.12; 2 ounces of insulation at $0.49; 2 ounces of gray fiber at $0.57; 1/2 yard of varnished cambric at $0.38; and labor costing $16.40?   _____

6. The thicknesses of insulation between the copper conductor and the iron core of a motor armature are 0.002 inch of enameled insulation, 0.006 inch of cotton insulation, 0.010 inch of slot insulation, and 0.008 inch of varnished cambric insulation. Find the total thickness of insulation.   _____

7. The costs of rewinding a motor are as follows: number 17 wire at a cost of $1.43; number 28 wire at a cost of $0.97; 1/2 pint of armalac at $1.89; and labor costing $18.60. Find the total cost of rewinding the motor.   _____

8. For a certain job, a contractor requires the following: BX cable, $32.65; conduit pipe, $12.56; toggle switches, $24.45; octal boxes, $7.50. What is the total cost of these materials?   _____

9. The total current is measured by the ammeter **A**. The total current in amperes is distributed in the circuit as shown. Find the total current reading on the ammeter.   _____

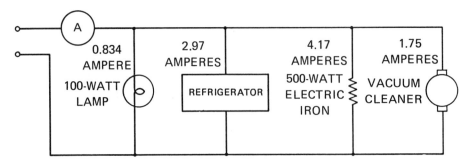

10. The values of pieces of electrical equipment are as follows: one 0-150 voltmeter costing $45.50; one 0-10-kilowatt meter costing $165.50; one 0-30 ammeter costing $35.20; one 0-300 voltmeter costing $42.75; one 0-200 millivoltmeter costing $65.38; one 50-ampere shunt costing $9.85; and three 0-5 ammeters costing $45.60 each. Find the total value of these pieces. _____

Note: Use this diagram for problems 11-15.

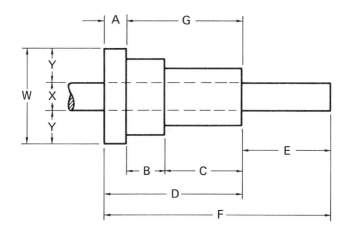

11. If A = 0.375 inch, B = 0.875 inch, and C = 1.062 5 inches, find D. _____

12. If A = 0.312 5 inch, B = 0.500 inch, and C = 1.062 5 inches, find D. _____

13. If A = 0.75 inch, B = 0.500 inch, C = 2.625 inches, and E = 3.125 inches, find F. _____

14. If A = 0.375 inch, B = 0.923 inch, and C = 1.099 2 inches, find G. _____

15. If X = 0.937 5 inch, and Y = 0.875 inch, find W. _____

16. The inside diameter of a motor shaft bearing is 0.007 5 centimetre larger than the shaft. The shaft diameter is 4.75 centimetres. What is the inside diameter of the bearing? _____

17. The inside diameter of the conduit shown is 4.035 centimetres. It is made of 0.3-centimetre thick steel. Find the outside diameter.

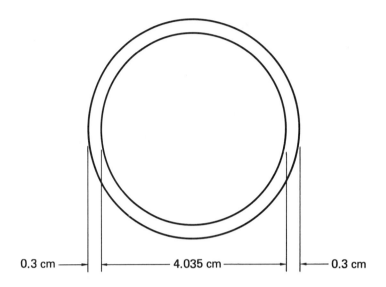

18. An electrician uses amounts of 1.27-centimetre EMT. The amounts used are 9.004 metres, 0.07 metre, 7.137 metres, 6.879 4 metres, and 578.9 metres. Find the total amount of EMT used.

Unit 12 SUBTRACTION OF DECIMAL FRACTIONS

BASIC PRINCIPLES OF SUBTRACTION OF DECIMAL FRACTIONS

When decimal fractions are subtracted, place the smaller number below the larger number taking care to keep the decimal points aligned. The procedure is then the same as subtraction of whole numbers.

Example: 12.349 – 8.907

$$\begin{array}{r} 12.349 \\ -8.907 \\ \hline 3.442 \end{array}$$

PRACTICAL PROBLEMS

1. A power plant generates 6 336.71 kilowatt-hours on Monday and 5 269.36 kilowatt-hours on Tuesday. How much less does the plant generate on Tuesday as compared to Monday? _____

2. The voltage at the terminals **A** and **B** in the diagram of the electric generator shown is 219.7 volts, and the voltage drop (loss in the line) is 3.96 volts. What is the voltage at the end of the line at points **C** and **D**? _____

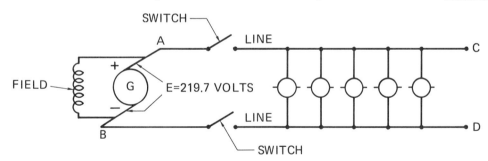

3. The total weight of a spool of number 16 wire is 11.01 pounds. The weight of the spool alone is 1.89 pounds. Find the weight of the wire only. _____

4. The cold resistance of an armature measures 0.096 ohm. After being run for 5 hours, the resistance is 1.25 ohms. What is the difference in the resistance? _____

5. A pin insulator measures 1.312 inches in diameter at the small end and 1.937 5 inches at the large end. What is the difference in diameters between the small end and the large end? _____

38 Section 3 Decimal Fractions

6. The field current of a motor is 1.12 amperes at no load. However, when the full load is applied, the current increases to 1.87 amperes. What is the difference in amperes from no load to full load? _____

7. A bearing bushing has an outside diameter of 3.937 5 inches and a wall thickness of 0.281 25 inch. Find the inside diameter of the bushing. _____

8. A special resistance wire has a diameter of 0.037 inch. The next smaller size is 0.034 5 inch in diameter. Find the difference in the diameters. _____

Note: Use this diagram for problems 9-13.

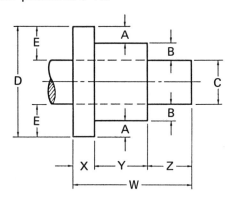

9. If D = 1.875 inches and E = 0.312 5 inch, find C. _____
10. If D = 3.187 5 inches and E = 0.562 5 inch, find C. _____
11. Find C if A = 0.375 inch, B = 0.25 inch, and D = 2.312 5 inches. _____
12. If W = 3.000 inches, Y = 1.750 inches, and Z = 0.875 inch, find X. _____
13. Find Z if X = 0.250 inch, Y = 2.062 5 inches, and W = 3.375 inches. _____

Note: Use this table for problems 14-17.

SOLID BARE COPPER CONDUCTORS

AMERICAN WIRE GAUGE SIZE NUMBER	WIRE DIAMETER IN INCHES
10	0.101 90
11	0.090 74
12	0.080 81
13	0.071 96
14	0.064 08
15	0.057 07
16	0.050 82

14. How much larger is the diameter of number 11 wire than of number 16 wire?

15. Is number 10 wire larger or smaller than number 12 wire?

16. What is the difference in diameters between number 13 wire and number 15 wire?

17. Sometimes in the manufacture of wire, the wire is slightly larger or smaller than intended. If a wire is measured and is found to be 0.079 inch in diameter, for what standard wire size is it intended?

18. A shim is made 0.002 5 inch thick. The thickness should be 0.002 25 inch thick. Find the difference in thicknesses.

19. A motor shaft 1 inch in diameter is worn down to 0.996 5 inch. How much is the diameter reduced by wear?

20. The total current in the diagram divides at point **A**. How much current exists at point **B**?

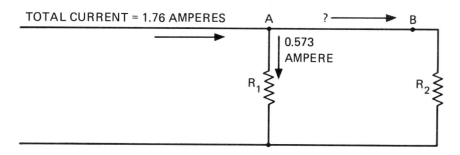

Unit 13 MULTIPLICATION OF DECIMAL FRACTIONS

BASIC PRINCIPLES OF MULTIPLICATION OF DECIMAL FRACTIONS

The multiplication of decimal fractions is the same basic procedure as the multiplication of whole numbers. When decimal fractions are multiplied, the number of places to the right of the decimal point in both numbers must be counted. The same number of places must appear to the right of the decimal point in the answer.

Example: 8.650 × 3.5

```
        8.650   (Three decimal places)
     X    3.5   (One decimal place)
        4 3250
       25 950
       30.2750  (Four decimal places)
```

PRACTICAL PROBLEMS

1. How much does it cost to operate a heater that uses 0.45 kilowatt of power if it is kept on for one hour? The cost is $0.14 per kilowatt-hour.

2. Find the total weight of a 52-gallon steel drum and its contents, if it contains 52 gallons of varnish weighing 8.16 pounds per gallon. The drum alone weighs 37.5 pounds.

3. The circumference of a circle equals the diameter × 3.1416. Find the circumference, to the nearer hundredth, of a pulley that has a diameter of 18.5 inches.

4. It is necessary to band an armature with 18 turns of steel wire per band. If the armature has a diameter of 20 inches, how many feet of wire are necessary per band, allowing 1.5 feet per band for waste and tightening?

5. A commutator has a diameter of 2.375 inches. What is the circumference, to the nearer thousandth, of the commutator?

6. If one cubic inch of copper weighs 0.314 pound, find the weight of a piece of copper that contains 2.25 cubic inches.

7. What is the weight of one gallon of sulphuric acid if 1 cubic inch weighs 0.066 5 pound? (231 cubic inches = 1 gallon)

8. What is the resistance of a piece of copper wire that has a size of 2.5 mil-feet?

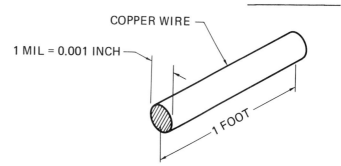

COPPER WIRE
1 MIL = 0.001 INCH
1 MIL × 1 FOOT = 1 MIL-FOOT
1 MIL-FOOT HAS A RESISTANCE OF 10.4 OHMS.
1 FOOT

9. An electric meter registers 8 749 kilowatt-hours on July 1, and on August 1 it registers 8 930 kilowatt-hours. The difference between these values is the total amount of energy used during this period. At $0.15 per kilowatt-hour, find the cost of the energy used.

10. What is the cost of 1 250 feet of cable at $23.25 per one hundred feet?

11. What is the cost of 1 750 feet of BX cable if it sells for $0.247 5 per foot?

12. What is the tax bill for a contractor if his shop has an assessed valuation of $35 000, and the tax rate is $27.50 per thousand dollars of assessed valuation?

13. A truck is known to consume about 0.12 gallon of gas per mile. If it is driven 18 500 miles and gas costs $0.939 per gallon, what is the cost for gas?

14. The circumference of a pulley equals the diameter × 3.141 6. Find the difference in circumferences between the two pulleys shown.

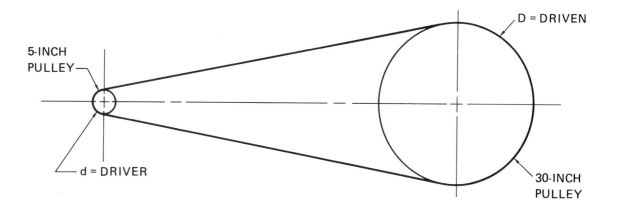

15. The drive pulley for the conveyer belt shown turns at the rate of 20 revolutions per minute. How many feet does the conveyer belt travel in one minute? (The distance for one revolution equals the circumference of the pulley.)

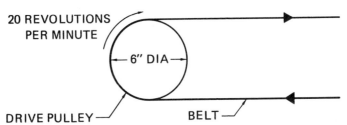

16. Six hundred twenty-five watertight screw connectors are purchased for $0.172 5 each. Find the total cost.

17. A dry cell delivers about 1.375 volts. What voltage will 27 dry cells deliver, if connected in series? (When connected in series, the voltage equals the sum of the voltages of the cells.)

18. A power company charges a city $1.216 7 per street lamp per month. There are 1 125 street lamps supplied with current. What is the city's power bill per year for the street lamps?

Unit 14 DIVISION OF DECIMAL FRACTIONS

BASIC PRINCIPLES OF DIVISION OF DECIMAL FRACTIONS

When decimal fractions are divided, the divisor is placed to the left of the dividend in the same manner as in the division of whole numbers. When dividing decimal fractions, however, the divisor must be a whole number and not a fraction. The divisor can be made a whole number by moving the decimal point all the way to the right of the number. When this is done, the decimal point of the dividend must be moved the same number of places to the right. The decimal point of the dividend is then placed directly above the division bracket. The numbers are then divided in the same manner as whole numbers.

Example: $19.44 \div 3.6$

```
          5.4
    3.6)19.4.4
         18 0
          1 4 4
          1 4 4
```

PRACTICAL PROBLEMS

1. If it costs $1 877 to construct 1/3 of a mile (5 280 feet = 1 mile) of an underground transmission system, what is the average cost of this job per foot? Express the answer to the nearer tenth of a cent.

2. The lamps in the figure require a total current of 7.76 amperes. How many amperes will 2 lamps require? (All the lamps are of the same type and each requires the same number of amperes.)

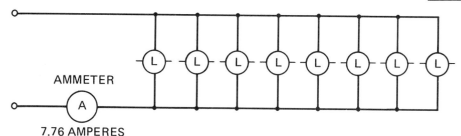

3. A certain size wire has a resistance of 2.56 ohms per one thousand feet. What is the resistance of 1 foot?

4. An electrician receives $434 for a 35-hour week. What is his rate of pay per hour?

44 Section 3 Decimal Fractions

5. If porcelain insulating tubes cost $10.56 per thousand, what is the loss to a school over a period of 27 school days if an average of 3 tubes are broken each day? Express the answer to the nearer cent. _____

6. A contractor pays an electrician $525 per week and a helper $367.50 per week. If the workweek consists of 35 hours, find the cost per hour for labor on a job using two electricians and two helpers. _____

7. The holes in the panel board shown are equally spaced across the board. Each hole is 1.25 inches in diameter, and the distance from each edge of the panel to the nearest hole is equal to the distance between holes. What is the length of the spaces between the holes? _____

28 INCHES

8. The weight of a spool of wire is 27.11 pounds. The spool weighs 3.2 pounds. Find the cost of the wire per pound if the cost of the spool of wire is $74.44. Express the answer to the nearer tenth of a cent. _____

9. The power for a circuit is 1 265.75 watts. Seven equal units use this amount of power. Find the number of watts, to the nearer hundredth, used per unit. _____

10. The lighting set shown has the same number of volts across each lamp. The sum of all lamp voltages is equal to the total number of volts. Find the number of volts across each lamp.

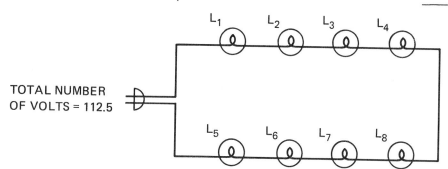

11. If 1 725 feet of Romex cable sell for $286.36, what is the cost of 1 136 feet of this cable?

12. The dry cells shown add up to a total voltage of 27.5 volts. What is the average number of volts for each dry cell?

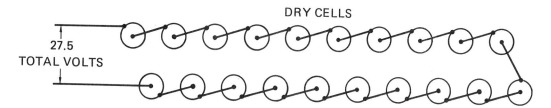

13. Three conductor, size 0000 cable weighs 3 530 pounds per one thousand feet. Find the weight of 475 feet of this cable.

14. The weight of a certain wire is 145 pounds per one thousand feet. What is the cost of 600 feet of this wire at $1.40 per one hundred pounds?

Unit 15 DECIMAL AND COMMON FRACTION EQUIVALENTS

BASIC PRINCIPLES OF DECIMAL AND COMMON FRACTION EQUIVALENTS

Changing Common Fractions to Decimal Fractions

A common fraction is an algebraic expression of division. It is possible to change any common fraction into its decimal equivalent by dividing the numerator by the denominator.

Example: Express 3/8 as a decimal.

$$\frac{3}{8} = 3 \div 8 = 0.375$$

Changing Decimal Fractions into Common Fractions

A decimal fraction is simply a common fraction which has a denominator that is some power of 10. The number of places to the right of the decimal point indicates the power of the denominator.

Example: Express 0.004 50 as a fraction.

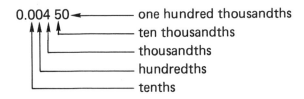

The fraction 0.004 50 has a denominator of 100 000. The decimal fraction is written as a common fraction and reduced to lowest terms.

$$\frac{450}{100\ 000} = \frac{9}{2\ 000}$$

PRACTICAL PROBLEMS

1. The armature shaft shown, originally 1 3/8 inches in diameter, is turned 1/16 inch smaller in diameter due to the irregular surface caused by running in a dry bearing. What is the finished size of the shaft, expressed as a decimal?

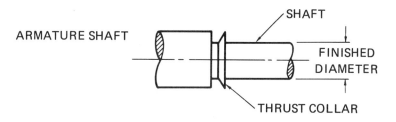

2. A piece of 1/4-inch diameter copper rod is rolled down in a mill to 0.010 inch in thickness for magnet wire sleeving and clips. In decimal form, how much is the copper reduced in size? _____

3. The total weight of a spool and the magnet wire on the spool is 29.14 pounds. If the spool weighs 3 3/4 pounds, find the weight of the wire. _____

4. Express the diameter of a 3/32-inch twist drill as a decimal. _____

5. If a machine screw measures 1/8 inch in diameter, what size hole must a washer have to be 0.006 inch larger than the screw? _____

6. A meter is to be bolted to a switchboard. The meter studs that will fit into the holes on the switchboard are 0.436 5 inch in diameter. Express the hole sizes in decimal form if they are to be 1/32 inch larger in diameter than the studs. _____

7. The motor bearing shown is to fit a shaft that measures 1.625 inches in diameter. What fractional size reamer should be used to ream a hole in the motor bearing? _____

8. A bearing with an inside diameter of 1 1/4 inches is found to be 0.008 inch oversize for the armature shaft. What should the diameter of the bearing be to fit the shaft? Allow 0.002-inch clearance for lubrication. _____

9. A piece of steel measures 1.375 inches in thickness before grinding and 1.250 inches after grinding. Express the amount of reduction in fractional form. _____

48 Section 3 Decimal Fractions

10. The carbon brush shown measures 23/32 inch thick and 1 15/32 inches wide. It is to have a copper plating on all sides 0.015 inch thick. What will be the finished dimension of the brush with the copper plating? _____

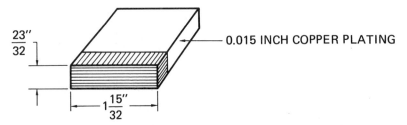

11. A bearing (babbitted) with a 1 3/4-inch bore is fitted to the new shaft of a motor by reaming. If the reaming process increases the diameter of the bore by 0.003 inch, what is the size of the bore after the work is completed? _____

12. A set of 2-inch wide by 5/8-inch thick carbon brushes fits the brush holders too tightly. It is necessary to take 5/1 000 inch off the width and 0.015 inch off the thickness by sanding. Give the width and thickness dimensions of the carbon brushes after sanding. _____

13. A steel sleeve bearing which has an outside diameter of 2 1/2 inches and an inside diameter of 2 1/8 inches must have a babbitt inserted to fit a shaft that has a diameter of 1.609 inches. What is the thickness of the babbitt shown? _____

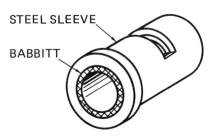

14. A piece of 0.062 5-inch thick copper is used to make a sleeve for a joint. What is the thickness of this material expressed as a fractional part of an inch? _____

15. The key shown in the figure is 7/16 inch wide and 5/8 inch deep. It is found that the key is too large in width (W) and depth (H). The width is machined off 0.005 inch and 0.003 inch is taken off the depth. What is the size of the key, expressed as decimals, after machining? _____

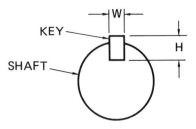

Unit 16 COMBINED OPERATIONS WITH DECIMAL FRACTIONS

BASIC PRINCIPLES OF COMBINED OPERATIONS WITH DECIMAL FRACTIONS

This unit provides practical problems involving combined operations of addition, subtraction, multiplication and division of decimal fractions.

PRACTICAL PROBLEMS

1. What is the total thickness of the following shims taken from a bearing: 0.065", 0.150", 0.130", 0.185", and 0.005"?

2. What is the total cost to repair an electric outlet, if the following materials and labor are required: one shallow box at $0.65, one 3/8-inch hickey at $0.98, one flush toggle switch at $1.17, and one hour of labor at $8.25?

3. What is the total number of amperes in a parallel circuit if the following lamps are connected to the circuit: one 100-watt lamp, 0.834 ampere; one 60-watt lamp, 0.437 ampere; one 40-watt lamp, 0.375 ampere; one 25-watt lamp, 0.225 ampere; one 10-watt lamp, 0.175 ampere; and one 7-watt lamp, 0.125 ampere?

4. What is the total thickness of insulation between the copper conductor and the iron core of a motor armature, if there is 0.003" of enameled insulation, 0.065" of cotton insulation, 0.015" slot insulation, and 0.007" of varnished cambric insulation?

Note: Use this illustration for problems 5-10.

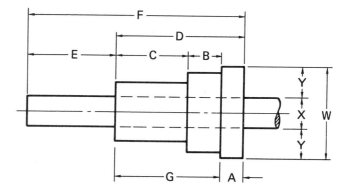

5. If A = 0.305", B = 0.870", C = 1.042 5", find D.
6. If X = 0.987 5" and Y = 0.675", find W.

7. If E = 3.165" and F = 7.025", find D.

8. Find C if B = 0.500" and G = 2.022 2".

9. If A = 0.305", B = 0.870", and D = 3.860", find G.

10. Find B if A = 0.305", C = 1.042 5", and D = 3.860".

11. If the total weight of a spool of number 16 wire is 15.625 pounds, and the weight of the spool alone is 7.25 pounds, find the weight of the wire.

12. The actual inside diameter of a 3" conduit is 3.375", and the actual outside diameter is 3.937 5". What is the wall thickness of this conduit?

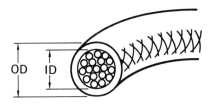

13. The field current of a motor is 7.25 amperes at no load. However, when the full load is applied, the current increases to 8.75 amperes. What is the difference in current from no load to full load?

14. What is the total weight of a container of 7 gallons of liquid insulating material, if one gallon weighs 9.36 pounds and the container alone weighs 6.87 pounds?

15. If the cost of Romex cable is $108.75 per one hundred feet, determine the total cost for 37 feet.

16. Determine the circumference of a grinding wheel if the diameter is 12 inches.

$$C = \pi D \quad \text{where} \quad \pi = 3.141\,6$$

17. An electrician works a total of 40 hours on a renovation job, and is paid a total of $340.00. Determine the hourly rate of pay.

18. A certain type of cable weighs 500.88 pounds per one thousand feet. Determine the weight of 53 feet of this cable. Round the answer to the nearer hundredth.

19. The holes drilled in the meter board shown are equally spaced across the board. The length of the board **L** is 56 inches, and the diameter of each hole is 2.75 inches. The distance from each edge of the panel to the nearest hole is the same as the distance **d** between the holes. Find **d**. Round the answer to the near hundredth.

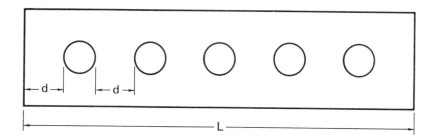

20. Standard 1-inch conduit has an inside diameter of 1.575 inches and an outside diameter of 1.825 inches. Find the thickness of the wall.

Percents, Averages, and Estimates

SECTION 4

Unit 17 PERCENT AND PERCENTAGE

BASIC PRINCIPLES OF PERCENT AND PERCENTAGE

Percent means hundredths or number per 100. If an electrician has 100 receptacle boxes and used 45 of them on a job, 45% of the supply is used on that job.

Changing a Decimal Fraction into a Percent

To change a decimal fraction into a percent, move the decimal point two places to the right and add a percent sign.

Example: Express 0.40 as a percent.

$$0.40 = 40\%$$

Changing a Common Fraction into a Percent

To change a common fraction into a percent, first change the common fraction into a decimal fraction by dividing the numerator by the denominator. Then move the decimal point two places to the right (multiply by 100) and add a percent sign.

Example: Express 5/8 as a percent.

$$\frac{5}{8} = 0.625 = 62.5\%$$

Changing a Percent into a Decimal Fraction

To change a percent into a decimal fraction, move the decimal point two places to the left (divide by 100) and drop the percent sign.

Example: Express 5% as a decimal fraction.

$$5\% = 0.05$$

Changing a Percent into a Common Fraction

To change a percent into a common fraction, first change the percent into a decimal fraction. Then change the decimal fraction into a common fraction and reduce to lowest terms.

Example: Express 12.5% as a common fraction.

$$12.5\% = 0.125 = \frac{125}{1\,000} = \frac{1}{8}$$

Finding What Percent One Number Is of Another

To find what percent one number is of another, first change the numbers into a fraction and then change the fraction into a percent.

Example: 40 is what percent of 50?

$$\frac{40}{50} = 0.80$$

$$0.80 = 80\%$$

PRACTICAL PROBLEMS

1. The generator shown ordinarily generates 1 500 volts. Find the percent of voltage increase that it is presently generating. _____

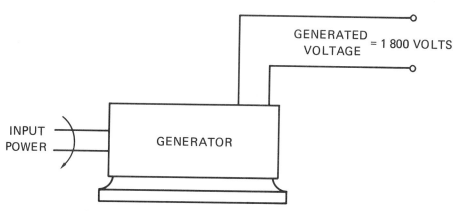

2. A motor rated at 90 horsepower is actually developing 105 horsepower. What is the percent of horsepower overload? _____

3. Each worker receives $43.20 per day. The wages are reduced 8%. Find, to the nearer cent, the amount each receives per day after the cut in pay. _____

4. In replacing 55 lamp bulbs, an apprentice breaks 6. Find, to the nearer hundredth percent, the percent broken. _____

5. Seven and one-half percent of a group of 640 motors are found to be overloaded. How many are not overloaded? _____

6. If $168.00 is the profit on a job and this represents 8% of the contract price, what is the contract price? _____

7. An electrical repairer charges 33% of the cost of a new motor for a rewinding job. If the motor costs $87.00 when new, what is the amount charged for rewinding? _____

8. Resistors R_1 and R_2 together use 63% of the total voltage. What is the voltage drop across resistor R_3? _____

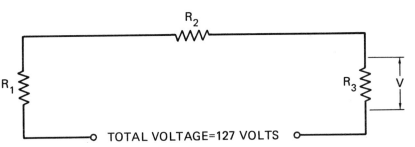

9. A motor that ordinarily delivers 40 horsepower is delivering only 36 horsepower. Find the percent that is now being delivered as compared to the usual amount. _____

10. A 12-volt battery had a capacity of 30 ampere-hours. Due to aging, the capacity drops to 24 ampere-hours. Find the percent decrease in capacities. _____

11. An electrician charges $175.00 for a wiring job. The cost of materials amounts to 62% of the total cost. Find the amount of money that the electrician receives for his labor. _____

12. Find, to the nearer hundredth horsepower, the amount of input horsepower required for the machine shown, if it is to deliver an output of 97 horsepower. _____

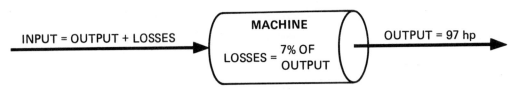

Unit 18 INTEREST

BASIC PRINCIPLES OF INTEREST

Interest is a way of using percentages. When money is borrowed, the borrowed amount is known as the *principal*. The amount charged for the use of the borrowed money is called the *interest* and the *rate* is expressed as a percent. Interest is generally computed on the basis of a 1 year period of time. When the term of the loan has expired, the money repaid is the sum of the principal and the interest.

Example: A company borrows $25 000 at an interest rate of 16% for a period of one year. How much money is repaid at the end of one year?

First, change the percent into a decimal fraction.

$$16\% = 0.16$$

Multiply the principal by the decimal fraction.

$$\$25\ 000 \times 0.16 = \$4\ 000$$

Add the interest charge to the principal to find the total amount of money repaid.

$$\$25\ 000 + \$4\ 000 = \$29\ 000$$

If the time of the loan is longer or shorter than one year, divide the annual interest by 12 months and multiply by the number of months. For example, assume the company was unable to repay the total loan at the end of 12 months and asked for a 6 month extension. Find the total amount of money repaid at the end of 18 months.

First, determine the interest for 12 months.

$$\$25\ 000 \times 0.16 = \$4\ 000$$

Divide the annual interest by 12 months.

$$\$4\ 000 \div 12 \text{ months} = \$333.33 \text{ per month}$$

Multiply the charge per month by the total number of months.

$$\$333.33 \times 18 = \$6\ 000$$

Add the total interest and principal to find the amount repaid.

$$\$25\ 000 + \$6\ 000 = \$31\ 000$$

PRACTICAL PROBLEMS

1. A contractor borrows $900.00 at 12% interest per annum. The debt is paid in 1 year and 7 months. Referring to the figure, find the amount paid back at the end of 1 year 7 months. _____

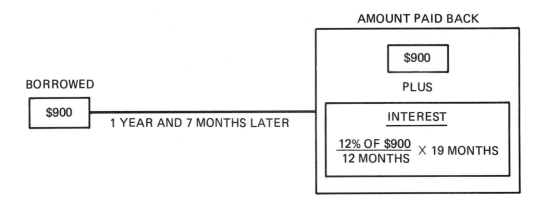

2. A contractor borrows $750.00 at 12% per annum and pays the debt six months later. What are the interest charges? _____

3. A contractor borrows $2 700.00 on July 1 at 12% per annum to purchase electrical supplies. If the loan is repaid 18 months later, how much interest is charged? _____

4. A contractor's note, given 2 years 9 months ago for $856.25 and bearing an 8% interest rate, per annum, is paid in full. What is the total amount of interest charges? _____

5. On June 1 five bills for electrical goods are sent out for the following amounts: $1 046.25, $952.40, $164.00, $1 150.00, and $518.00. On December 1 of the same year, payment is received in full, with interest at the rate of 12% per annum. What is the total amount received? _____

6. An electrician borrows $1 250.00, at 12.5% per annum, to replace direct current motors with alternating current motors. The loan is paid back in 1 year and 2 months. Find the amount of interest. _____

7. A firm purchases heavy cable and 4-inch conduit on credit, and agrees to pay 10% interest per annum. Purchases are made in October for cable costing $756.80 and in December for conduit costing $1 325.25. If full payment is made by March 1st of the following year, how much is paid? (Charge interest for the full month of purchase.) _____

8. A firm purchases several motors at a cost of $8 350.00 and gives a note bearing an 8% interest rate per annum. The note is due one year later. What is the total amount to be paid on the note? _____

9. On July 17, a contractor borrows $1 150.00 at an interest rate of 10% per annum, and pays his note when it becomes due 39 months later. What is the amount of interest? _____

10. An electrician purchases a used truck for the amount shown. He trades in an old truck, and pays an amount of money in cash toward the total payment. When the loan is paid in full after one year, what will it cost in interest payments? _____

BILL OF SALE	
Cost of Used Truck	$821.00
Trade-in Allowance	75.00
Cash Payment	250.00
Balance Due	?
Interest On Balance	12% Per Annum
Time to Pay in Full	1 Year

11. A contractor has $650.00 worth of electrical material on the shelf in stock for 3 consecutive years. If the money is borrowed at 12% per annum, what does it cost to carry this stock? _____

Unit 19 DISCOUNT

BASIC PRINCIPLES OF DISCOUNT

Discounts are another way of using percentages. A discount is subtracted from the list price to find the net price.

Example: A crimp connector tool lists for $18.50. The tool is offered at a 12% discount. What is the net price of the tool?

First, change 12% into a decimal fraction.

$$12\% = 0.12$$

Multiply the list price by the discount.

$$\$18.50 \times 0.12 = \$2.22$$

Subtract the discount from the list price.

$$\$18.50$$
$$-\ \ 2.22$$
$$\$16.28 \text{ (Net price of the tool)}$$

PRACTICAL PROBLEMS

1. An electric pump is listed at $39.50. Find the net cost of the pump with a 20% discount. _____

2. A contractor purchases quantities of wire, fittings, and switches listed at $2 150.00, and receives discounts of 15%, 10%, and 3%. What is the net cost? _____

MULTIPLE DISCOUNT

LIST PRICE − (0.15 × LIST PRICE) = FIRST DISCOUNT PRICE

FIRST DISCOUNT PRICE − (0.10 × FIRST DISCOUNT PRICE) = SECOND DISCOUNT PRICE

SECOND DISCOUNT PRICE − (0.03 × SECOND DISCOUNT PRICE) = NET COST

3. Two-inch conduit is listed at $161.56 per hundred feet. An electrician purchases 625 feet of conduit and receives trade discounts of 20%, 5%, and 2%. What is the net cost? _____

4. Solderless lugs are quoted at $42.50 less 40% for standard packages of 50. What will be the cost if only 22 are ordered at list price less 35%? _____

5. Universal 2 1/2-inch conduit caps are listed at $7.00 each, less 20%, 15%, and 2%. Sixteen of these are ordered. What is the total net cost? _____

6. A contractor purchases motors listed at $6 800.00 and receives discounts of 20% and 10%. What is the net cost? _____

7. An electrician purchases 1 250 feet of two-inch flexible steel conduit at a list price of $109 per hundred feet. With discounts of 20% and 10%, what is the net cost? _____

8. Three electrical distributors offer the same grade of materials at the same list price. Distributor **A** offers discounts of 25% and 15%; distributor **B** offers discounts of 20% and 20%; distributor **C** offers discounts of 15%, 15%, and 10%. Materials listed at $350 are to be purchased.

   a. Find the net cost when the materials are purchased from distributor **A**. a. _____

   b. Find the net cost when the materials are purchased from distributor **B**. b. _____

   c. Find the net cost when the materials are purchased from distributor **C**. c. _____

9. Find the total net cost of the shipment of materials shown. _____

ITEM	QUANTITY	LIST	DISCOUNT
Floor Box	10	$4.00 Each	20%
1/2-inch Condulet	250	$2.26 Each	25%
Single Pole Key Socket Body	250	$4.14 Each	40%
Single Pole Flush Switch	35	$1.27 Each	40%

10. What is the net price of a 2-pole, 400-ampere, 230-volt, enclosed-entrance switch if it has a list price of $287.00, with discounts of 48% and 2%? _____

11. Seventy-two 30-ampere, 600-volt, enclosed, refillable fuse cases are purchased at a list price of $27.50 each with a discount of 28%. What is the total cost? _____

12. An electrician purchases thirty 125-volt, 30-ampere, double-pole, double-branch cutouts listed at $6.40 per box of 5, less 25%; and eight surface panels listed at $4.75 each, less 35%. Three percent is saved by paying the bill in 15 days. What is the cost if paid within 15 days? _____

Unit 20 AVERAGES AND ESTIMATES

BASIC PRINCIPLES OF AVERAGES AND ESTIMATES

Averages

Averages are found by adding each individual amount of some number of units and dividing the sum by the total number of units.

Example: A company truck has gasoline expenses of:

First week	$68.50
Second week	$87.34
Third week	$103.09
Fourth week	$77.16

What is the average weekly gas bill for this truck?

Find the sum of the total gasoline bills for the month, and then divide the total by the number of weeks.

```
    68.50
    87.34
   103.09
 + 77.16
   336.09
```
$336.09 \div 4 = \$84.02$ (average)

Estimates

Estimates are used for approximation and are not intended to be exact. Being able to estimate the cost of materials for a job or the amount of time needed to perform a certain task requires time and experience.

Example: An electrician estimates the materials for a certain job will cost $120.00. When the job is finished, it is found that the actual cost of materials is $112.16. The actual amount is within what percent of the estimate?

$$\$112.16 \div \$120 = 93.5\% \qquad 100\% - 93.5\% = 6.5\%$$

The actual amount is within 6.5% of the estimated amount.

PRACTICAL PROBLEMS

1. The amounts of power used in an electrical maintenance shop are as follows: April, 41.2 kilowatt-hours; May, 59.25 kilowatt-hours; June, 53.63 kilowatt-hours; July, 62.4 kilowatt-hours; August, 63.75 kilowatt-hours; September, 30.35 kilowatt-hours. What is the average monthly power usage? Express the answer to the nearer hundredth.   _____

2. A 208-volt, 4-wire, 3-phase wye power system feeds a school building. Daily voltage readings taken for one week are 205.25 volts, 203.75 volts, 208 volts, 204.35 volts, 206.7 volts, 207 volts, and 208.55 volts. What is the average daily voltage of the system? Round the answer to the nearer hundredth.   _____

3. An electrician earns $10.75 per hour. During one week the electrician works these hours: Monday, 8 hours; Tuesday, 7 hours; Wednesday, 5 1/2 hours; Thursday, 10 hours; and Friday, 4 1/2 hours. What is the average daily earning?   _____

4. An electrician estimates 3 1/2 bundles (approximately 106.68 metres) of 1/2-inch conduit are needed to wire a house. Each bundle costs $47.75 or $1.57 per metre. The amounts used are as follows: living room, 13.75 metres; dining room, 22.6 metres; kitchen, 28.5 metres; bedroom A, 12.1 metres; bedroom B, 12.3 metres; and bathroom, 6.8 metres.
   a. What is the average cost per room?   a. _____
   b. The electrician's estimate of the cost is how much over or under the cost of the job?   b. _____

5. Five light fixtures for a house cost these amounts: type A, $15.83; type B, $21.75; type C, $19.69; type D, $24.30; type E, $9.49.
   a. What is the average cost per fixture?   a. _____
   b. The electrician estimates a cost of $19.00 per fixture. How much over or under is the estimate for the five fixtures?   b. _____

6. Two-inch angle iron weighing approximately 2.97 kilograms per metre is used to build panel mounting racks. These amounts of iron are used: panel A, 9 metres; panel B, 10.75 metres; panel C, 8.6 metres; panel D, 7.7 metres. What is the average weight per panel? Round the answer to the nearer hundredth.   _____

62 Section 4 Percents, Averages, and Estimates

7. An electrician estimates 760 metres of No. 12 NM cable needed to wire a house. Each coil of cable contains 76.2 metres. The amounts used in different rooms are 111.8 metres, 97.9 metres, 401.7 metres, and 112.5 metres.
   a. How many coils of wire are used?   a. _____
   b. How many metres over or under is the estimate?   b. _____

8. A shipment of 15 heat pumps arrive for storage at an electrical supply house. The pumps cannot be stacked and each has a base measurement of 3.2 metres x 3.8 metres. A storage area 14 metres long and 12 metres wide is available. Is there sufficient space to store the shipment?   _____

9. The temperature of a walk-in cold provisions cabinet is sampled twice daily for a 3-day period. The Fahrenheit temperature readings are 37°, 33°, 35.5°, 34°, 38°, and 36.6°. What is the average temperature? Express the answer to the nearer tenth.   _____

10. It is required that a cold storage temperature be maintained at an average of 3.055 degrees Celsius plus or minus 35%.
    a. Find the lowest temperature reading that satisfies this requirement. Express the answer to the nearer thousandth.   a. _____
    b. Find the highest temperature reading that satisfies this requirement. Express the answer to the nearer thousandth.   b. _____

Unit 21 COMBINED PROBLEMS ON PERCENTS, AVERAGES, AND ESTIMATES

BASIC PRINCIPLES OF PERCENTS, AVERAGES, AND ESTIMATES

This unit provides practical problems involving combined problems on percents, averages, and estimates.

PRACTICAL PROBLEMS

1. A motor rated at 30 horsepower is actually developing 35 horsepower. What is the percent of horsepower overload? Round to the nearer tenth.

2. Workers receiving $43.20 per day have their wages increased 8%. How much is received per day after the increase?

3. An engine has an input of 90 horsepower and uses 8% of its input power to overcome friction and other losses. Find the available horsepower at the output if the output = input − losses.

4. When mixing a quantity of electrolyte for a storage battery, the electrician uses 2 parts of acid and 3 parts of water. What percent is acid?

5. A contractor borrows $1 000, at 11.8% per annum, to buy material for a job. The debt is paid 18 months later. What amount is paid in interest?

6. An electrician borrows $2 750.00 at 9.7% per annum, to purchase electrical supplies. If the loan is repaid in 15 months, how much does it cost for interest?

7. A contractor borrowed $3 500 at an interest rate of 6% per annum, and pays the note when it became due 6 months later. What is the amount of interest paid for the use of the money?

8. A contractor purchases a truck for $8 350 and is allowed $2 000 for trading-in an old truck. A cash deposit of $200 is made. How much interest is paid for one year at a rate of 12% per annum?

9. Find the total net cost of the following shipment of materials: 23 floor boxes, listed at $4.00 each, less 15%; 150 half-inch Condulets, listed at $0.60 each, less 20%; 125 convenience outlets, listed at $0.79 each, less 10%; 83 single-pole flush switches, listed at $0.45 each, less 30%.

Section 4 Percents, Averages, and Estimates

10. What is the net price of a 2-pole, 100-ampere, 230-volt entrance switch, if the list price is $137.00 with discounts of 35% and 3%? _____

11. An electrician is paid $1 420 in January, $1 560 in February, $1 672 in March, $1 878 in April, $1 925 in May, and $2 016 in June. What is the average monthly pay? _____

12. An electrical supply uses the following watts: 4 212, 4 226, 4 296, 3 427, 914, 1 428, and 4 293. What is the average wattage used? Round the answer to the nearer hundredth. _____

13. An electrician estimates using 500 staples on a six-room house. The electrician uses 70 staples in the first room, 80 staples in the second room, 50 staples in the third room, 75 staples in the fourth room, 95 staples in the fifth room, and 100 staples in the sixth room. Find the percent of accuracy for the estimate. _____

14. Find the average for the following measurements: 17 millimetres, 16 millimetres, 18 millimetres, 21 millimetres, 29 millimetres, and 24 millimetres. Round the answer to the nearer hundredth. _____

Powers and Roots

SECTION 5

Unit 22 POWERS

BASIC PRINCIPLES OF POWERS

The power of a number is also known as its *exponent*. When a number is raised to a certain power, the number is used as a factor (number in multiplication) that number of times. If a number is squared, for example, it is raised to the second power, or has an exponent of 2. This means that the number is to be multiplied times itself. The exponent is a small number written above and to the right of the number as shown.

$$5^2 = (5 \times 5) = 25$$

Notice that five is not multiplied by 2, but rather is multiplied by itself.

If a number is cubed, it means that the number is raised to the third power or has an exponent of 3.

$$8^3 = (8 \times 8 \times 8) = 512$$

A number can be raised to any power by using it as a factor that number of times.

Example: Raise 6 to the fifth power.

$$6^5 = (6 \times 6 \times 6 \times 6 \times 6) = 7\,776$$

PRACTICAL PROBLEMS

Raise the following expressions to the power indicated:

1. 7^2 _____
2. 8^2 _____
3. 9^2 _____
4. 10^2 _____
5. 11^2 _____
6. 7^3 _____
7. 8^3 _____
8. 9^4 _____
9. 10^4 _____
10. 11^5 _____
11. 15^2 _____
12. 21^2 _____
13. 25^3 _____
14. 28^3 _____
15. 30^4 _____
16. 35^5 _____
17. 43^5 _____
18. 50^6 _____

Note: For a round conductor, the diameter in *mils* squared (d^2) equals the cross-sectional area in *circular mils* (CM). In other words, d^2 = the circular mil area.

1 mil = 0.001 inch

65

19. Find the cross-sectional area in circular mils for the wire shown. _____

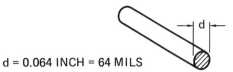

d = 0.064 INCH = 64 MILS

20. Find the cross-sectional area, in circular mils, of a wire that is 104 mils in diameter. _____

21. What is the cross-sectional area, in circular mils, of a conductor that has a diameter of 32 mils? _____

Note: The symbols used in problems 22-25 are as follows:

P = Watts (power)   I = Amperes (current)
E = Volts (voltage)   R = Ohms (resistance)

22. To find the number of watts used in any circuit, the amperes squared are multiplied by ohms. Find the number of watts for the device shown. _____

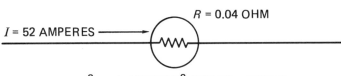

$P = I^2 R = (AMPERES)^2 (OHMS) = WATTS$

23. Find the wattage for the circuit shown. _____

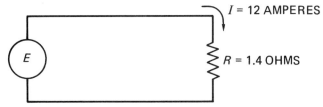

Note: Use this illustration for problems 24-25.

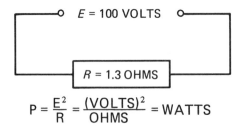

$P = \dfrac{E^2}{R} = \dfrac{(VOLTS)^2}{OHMS} = WATTS$

24. The number of watts used in a circuit is equal to the volts squared divided by the ohms. Find the wattage used in the circuit shown. Round the answer to the nearer tenth. _____

25. If the voltage is changed to 220 volts, and the resistance is changed to 3.6 ohms, how many watts are used? Round the answer to the nearer tenth. _____

26. A cable is checked for a power installation and is found to be too small. The cable has 37 strands of copper wire, each having a diameter of 82.2 mils. Find the size of the cable in circular mils. _____

27. If a flexible copper cable has 133 strands, each 5.63 mils in diameter, what is the size of the cable in circular mils? _____

Unit 23 ROOTS

BASIC PRINCIPLES OF ROOTS

The root of a number is the opposite of its power. The root of a number is determined by finding what number used as a factor some number of times will equal that number. The square root of 36 is 6. 6 is the square root of 36 because 6 squared equals 36.

$$(6 \times 6) = 36$$

This is true for other roots also. The cube root of 27 is 3, because 3 cubed equals 27.

$$(3 \times 3 \times 3) = 27$$

When the root of a number is to be found, the number is placed under a radical sign ($\sqrt{}$). If the square root of a number is to be found, no number is used outside of the radical sign.

$$\sqrt{25} = 5$$

If some root other than a square root is to be found, a small number is placed outside the radical sign. This number is used to indicate what root is to be found.

$$\sqrt[3]{64} = 4$$

Methods of Finding Square Roots

There are several methods which can be used to find the square root of a number. Some of these include square root charts, calculators, or the long hand method. In the following example, the long hand method for finding a square root is shown.

Example: Find the square root of 1 800.

In this example the square root of 1 800 will be found. The first step is to divide the number into groups of two numbers each, left and right of the decimal point.

$$\sqrt{18\ 00\ .\ 00\ 00}$$

Considering only the first two numbers in the group, find the smallest number that can be squared without going over the first two numbers. In this case, the number is 4, because $4 \times 4 = 16$. Place the 16 below the 18 and subtract. Also place 4 beside 16.

$$\begin{array}{r} 4 \\ \sqrt{18\ 00\ .\ 00\ 00} \\ 4\ |\ 16 \\ \hline 2 \end{array}$$

Bring down the next two numbers beside the 2. Double the answer (4) and place it beside the problem shown. Now find a number that can be placed beside the 8 so that when that number is multiplied times the combined number it will be closest to but not go over 200. In this problem the number will be 2. When the 2 is placed beside the 8, it becomes 82. 82 × 2 = 164. 164 is then placed below the 200. Subtract 164 from 200.

```
            4  2.
         _____
       √ 18 00 . 00 00
      4|  16
          ────
           2 00
      82|  1 64
          ─────
             36
```

Bring down the next two numbers beside the 36. Double the answer and place this answer beside the problem. (42 × 2 = 84). Find a number when added to 84 and then multiplied by that number will not go over 3 600. In this problem, that number is 4. 844 × 4 = 3 376. Subtract 3 376 from 3 600.

```
            4  2. 4
         _____
       √ 18 00 . 00 00
      4|  16
          ────
           2 00
      82|  1 64
          ─────
             36  00
     844|    33  76
          ───────
              2 24
```

Bring down the next set of numbers beside 224. Double the answer and place this number beside the problem. (424 × 2 = 848). Find a number that can be added to 848 that when it is multiplied by that number will not go over 22 400. In this problem, that number is 2. (8482 × 2 = 16 964). Subtract 16 964 from 22 400. Follow this procedure until the answer contains as many decimal places as desired.

```
            4  2. 4  2
         _____
       √ 18 00 . 00 00
      4|  16
          ────
           2 00
      82|  1 64
          ─────
             36  00
     844|    33  76
          ───────
              2 24 00
    8482|    1 69 64
          ──────────
                54 36
```

Section 5 Powers and Roots

PRACTICAL PROBLEMS

Solve the following expressions and express the answer to the nearer hundredth.

1. $\sqrt{81}$ _____
2. $\sqrt{100}$ _____
3. $\sqrt{169}$ _____
4. $\sqrt{361}$ _____
5. $\sqrt{529}$ _____

6. $\sqrt{743}$ _____
7. $\sqrt{892}$ _____
8. $\sqrt{1\,235}$ _____
9. $\sqrt{1\,692}$ _____
10. $\sqrt{2\,000}$ _____

Note:
- P = Watts (power)
- E = Volts (voltage)
- I = Amperes (current)
- R = Ohms (resistance)

11. The circuit shown uses 2 048 watts. How many amperes are in the circuit? _____

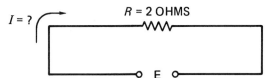

$$\text{CURRENT} = I = \sqrt{\frac{P}{R}} = \sqrt{\frac{\text{WATTS}}{\text{OHMS}}} = \text{AMPERES}$$

12. An electrical cable has a cross-sectional area of 1 000 000 circular mils, and contains 19 strands. Find the diameter of each strand. _____

$$\text{Diameter} = \sqrt{\text{number of circular mils per strand}}$$

13. An electric heating device uses power in the amount of 605 watts. The resistance of the device is 20 ohms. This heater operates on what voltage? _____

$$E = \sqrt{P \times R}$$

Note: Use the circuit and the associated formulas shown for problems 14-22.

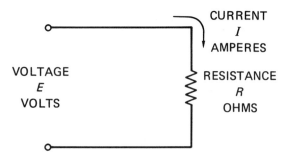

POWER = WATTS

$$\text{AMPERES} = \sqrt{\frac{\text{WATTS}}{\text{OHMS}}}$$

$$\text{VOLTS} = \sqrt{\text{WATTS} \times \text{OHMS}}$$

14. If the resistance has a value of 550 ohms and uses 22 watts of power, find the number of amperes that the resistance takes. _____

Unit 23 Roots 71

15. Eight hundred eighty watts with 220 ohms of resistance requires what value of current in amperes? _____

16. What voltage in volts is required if the power is 25 watts and the resistance is 64 ohms? _____

17. What current in amperes is present in the circuit if the resistance is 0.88 ohm and the power is 13.75 watts? _____

18. When the power is 220 watts and the resistance is 100 ohms, what voltage is in the circuit? _____

19. When the resistance is 495 ohms and the power is 22 watts, what value of current exists in amperes? _____

20. When the resistance is 73.3 ohms and the power is 165 watts, what is the value of the current in amperes? _____

21. The resistance uses 287 watts and has a value of 46 ohms. How many amperes exist in the circuit? _____

22. Find the number of volts in the circuit if the resistance has a value of 22 ohms, and uses 550 watts when the current value is 5 amperes. _____

23. The cross-sectional area in circular mils of a number 10 wire is 10 380. The diameter of a wire, in mils, is found by taking the square root of the circular mil area. Find the diameter of the number 10 wire. _____

24. If a 250 000-circular mil cable is composed of 37 strands, what is the approximate diameter of each strand? _____

25. To find the line voltage in the 2-phase, 3-wire circuit shown, the constant $\sqrt{2}$ is used. The line voltage is equal to $\sqrt{2}$ times E ($E_{LINE} = \sqrt{2} \times E$). Find the value of line voltage. _____

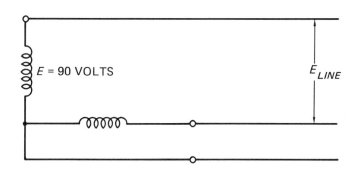

E = 90 VOLTS

E_{LINE}

72 Section 5 Powers and Roots

26. A 5 000 000-circular mil cable is composed of 37 strands of copper wire of equal size. What is the approximate diameter of one strand? _____

27. The square root of 3 is used to make calculations for 3-phase, 3-wire circuits like the one shown. The line current for this type of circuit is equal to $\sqrt{3}$ times I ($I_{LINE} = \sqrt{3} \times I$). Find the line current for this circuit. _____

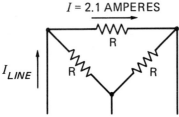

I = 2.1 AMPERES

I_{LINE}

28. If the cross-sectional area in circular mils of a wire is 4.107, what is the diameter? _____

29. A 750 000 circular mil cable is composed of 61 strands of copper wire of equal size. What is the approximate diameter of one of the strands? _____

Unit 24 COMBINED OPERATIONS WITH POWERS AND ROOTS

BASIC PRINCIPLES OF COMBINED OPERATIONS WITH POWERS AND ROOTS

This unit provides practical problems involving combined operations with powers and roots.

PRACTICAL PROBLEMS

1. $\sqrt{121}$ _____
2. $\sqrt{196}$ _____
3. $\sqrt{289}$ _____
4. $\sqrt{441}$ _____
5. $\sqrt{729}$ _____
6. $\sqrt{1\,156}$ _____
7. $\sqrt{1\,521}$ _____
8. $\sqrt{1\,936}$ _____
9. $\sqrt{3\,136}$ _____
10. $\sqrt{4\,356}$ _____

11. The diameter of #10 copper wire is 101.9 mils. Find the cross-sectional area of the wire in circular mils.   _____

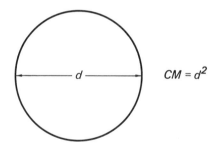

$CM = d^2$

12. A cable is checked for a power job and is found to be too small. The cable has 62 strands of copper wire having a diameter of 37.2 mils each. What is the size of the cable in circular mils?   _____

13. The cross-sectional area in circular mils of a number 12 wire is 6 530. Find the diameter. Round the answer to the nearer mil.   _____

14. A 2 583 circular mil cable is composed of 25 strands of copper wire of equal size. What is the approximate diameter of one strand? Round the answer to the nearer tenth.   _____

15. A circuit has a power rating of 500 watts and a resistance of 65 ohms. What is the current flow in the circuit? Round the answer to the nearer hundredth of an ampere.   _____

$$I = \sqrt{\frac{P}{R}} \quad ; \quad \text{amperes} = \sqrt{\frac{\text{watts}}{\text{ohms}}}$$

Section 5 Powers and Roots

16. The resistance of a coil of wire that uses 800 watts is 62 ohms. What is the voltage of the circuit? Round the answer to the nearer hundredth.

$$E = \sqrt{P \times R} \; ; \qquad \text{volts} = \sqrt{\text{watts} \times \text{ohms}}$$

17. If a circuit uses 700 watts and has a resistance of 50 ohms, what is the amperage of the circuit? Round the answer to the nearer hundredth.

$$I = \sqrt{\frac{P}{R}}$$

Measure

SECTION 6

 Unit 25 LENGTH MEASURE

BASIC PRINCIPLES OF LENGTH MEASURE

The English System of Measure

The English system of measure was originally based on the measurements of the king's body and other natural things. The length of the king's foot, for example, was the standard length of a foot. The middle joint of the king's index finger was the standard length of one inch. Another weight measurement still used today is the grain. A grain was the weight of a grain of wheat.

These measurements are impossible to duplicate, because no two things that occur in nature are exactly the same. No two grains of wheat weigh exactly the same, for instance. Today, there are standards of measurement that are kept by the bureau of standards. The length of a foot, for example, is always the same.

The Metric System

The standard length in the metric system is the meter. The length of the meter was originally based on the measurement of the earth. Today, its standard is the length of ray emitted by the element krypton. The metric system is also based on a standard value of 10. This is an advantage over the English system. Lengths of English and metric measure are shown in the following charts.

ENGLISH LENGTH MEASURE	
12 inches (in)	= 1 foot
3 feet (ft)	= 1 yard (yd)
1 760 yards (yd)	= 1 mile (mi)
5 280 feet (ft)	= 1 mile (mi)

METRIC LENGTH MEASURE	
10 millimetres (mm)	= 1 centimetre (cm)
10 centimetres (cm)	= 1 decimetre (dm)
10 decimetres (dm)	= 1 metre (m)
10 metres (m)	= 1 dekametre (dam)
10 dekametres (dam)	= 1 kilometre (km)

ENGLISH-METRIC EQUIVALENTS LENGTH MEASURE	
1 inch (in)	= 25.4 millimetres (mm)
1 inch (in)	= 2.54 centimetres (cm)
1 foot (ft)	= 0.304 8 metre (m)
1 yard (yd)	= 0.914 4 metre (m)
1 mile (mi)	≈ 1.609 kilometres (km)
1 millimetre (mm)	≈ 0.039 37 inch (in)
1 centimetre (cm)	≈ 0.393 70 inch (in)
1 metre (m)	≈ 3.280 84 feet (ft)
1 metre (m)	≈ 1.093 61 yards (yd)
1 kilometre (km)	≈ 0.621 37 mile (mi)

76 Section 6 Measure

PRACTICAL PROBLEMS

1. Find the number of millimetres in 1.2 metres.   _____
2. Find the number of centimetres in 54 millimetres.   _____
3. Find the number of millimetres in 3.7 centimetres.   _____
4. Find the number of metres in 4 376 millimetres.   _____
5. Find the number of metres in 43 centimetres.   _____
6. Express 3/4 inch to the nearer hundredth millimetre.   _____
7. Express 6 inches to the nearer hundredth centimetre.   _____
8. Express 9 feet to the nearer thousandth metre.   _____
9. Express 55 miles to the nearer thousandth kilometre.   _____
10. Express 100 yards to the nearer hundredth metre.   _____
11. A wiring job requires 73 metres of 3.8-centimetre EMT. There are 9 pieces of the required EMT on hand. Each piece on hand measures 20 feet. How many additional 20-foot lengths need to be ordered for this job?   _____
12. The outside diameter of an electrical conduit is 6.3 centimetres. The thickness of the conduit is 0.4 centimetre. Find the inside diameter of the conduit.   _____

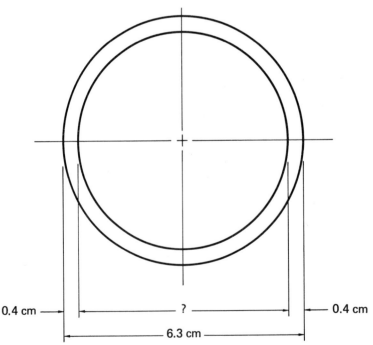

Note: Use this illustration for problems 13 and 14.

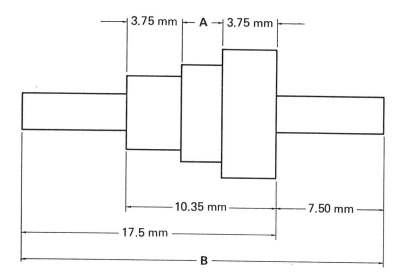

13. Find dimension **A** in millimetres.
14. Find dimension **B** in centimetres.
15. A concrete pad 5 metres by 7 metres is poured adjacent to a wall to support a light dimmer. The dimmer is 91 centimetres wide, 457 centimetres long, and 366 centimetres high. The dimmer is placed with a 1 metre clearance from the wall. What is the distance, in metres, across the width of the pad from the base of the dimmer to the edge of the pad?

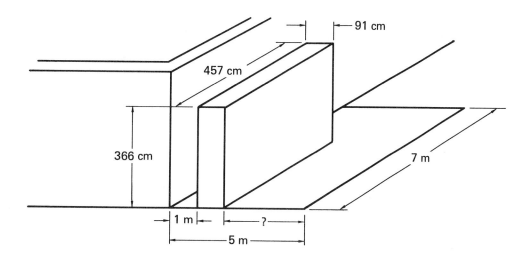

78   Section 6 Measure

16. To run a 3-wire cable, a hole is drilled through a plate, the subflooring, and the finished flooring. The plate is 1.6 centimetres and the subflooring is 3.4 centimetres. The finished flooring is 8 millimetres. Find the depth of the hole in centimetres. _____

17. A hole is cut in a Sheetrock wall for installation of a junction box that is 7.62 centimetres wide and 12.7 centimetres long. A clearance of 0.3 centimetre is required on all sides of the box.

   a. What is the width of the hole?   a. _____

   b. What is the length of the hole?   b. _____

18. Find the distance A in the eccentric cam shown.   _____

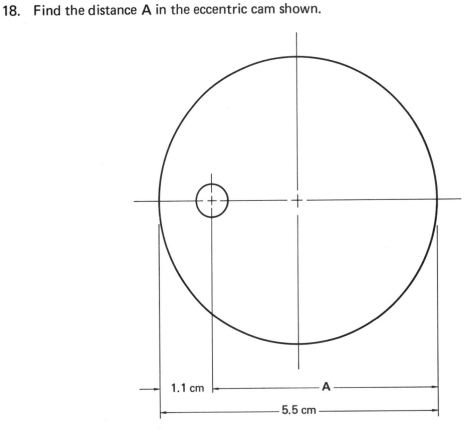

Unit 26 AREA MEASURE

BASIC PRINCIPLES OF AREA MEASURE

The charts that follow show area measurements for both the English and metric systems. The measurement of area is a two dimensional quantity. The measurement of area is found by multiplying length and width. A good example of this is the size of the image projected by a slide projector. The length and width of the image can be measured.

ENGLISH AREA MEASURE	
144 square inches (sq in)	= 1 square foot (sq ft)
9 square feet (sq ft)	= 1 square yard (sq yd)

METRIC AREA MEASURE	
100 square millimetres (mm^2)	= 1 square centimetre (cm^2)
100 square centimetres (cm^2)	= 1 square decimetre (dm^2)
100 square decimetres (dm^2)	= 1 square metre (m^2)
100 square metres (m^2)	= 1 square dekametre (dam^2)
100 square dekametres (dam^2)	= 1 square hectometre (hm^2)
100 square hectometres (hm^2)	= 1 square kilometre (km^2)

ENGLISH-METRIC EQUIVALENTS AREA MEASURE	
1 square inch (sq in)	= 645.16 square millimetres (mm^2)
1 square inch (sq in)	= 6.451 6 square centimetres (cm^2)
1 square foot (sq ft)	≈ 0.092 903 square metre (m^2)
1 square yard (sq yd)	≈ 0.836 127 square metre (m^2)
1 square millimetre (mm^2)	≈ 0.001 550 square inch (sq in)
1 square centimetre (cm^2)	≈ 0.155 00 square inch (sq in)
1 square metre (m^2)	≈ 10.763 910 square feet (sq ft)
1 square metre (m^2)	≈ 1.119 599 square yards (sq yd)

Section 6 Measure

- Study these formulas of area measure.

Areas: (A = area)

Circle $\pi = 3.1416$
$A = \pi r^2$ r = radius
$A = \frac{\pi}{4} D^2$ D = diameter
$\frac{\pi}{4} = 0.7854$

Rectangle l = length
$A = lw$ w = width

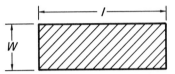

Square s = length of side
$A = s^2$

Trapezoid B = length of large side
$A = \frac{(B + b)\,a}{2}$ b = length of small side
a = altitude (height)

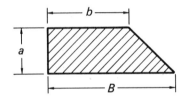

Triangle a = altitude
$A = \frac{ab}{2}$ b = base

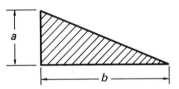

PRACTICAL PROBLEMS

1. Express 0.75 square metre as square millimetres.

2. Express 22 575 square millimetres as square metres.

3. Express 360 square inches as square feet.

4. Find, to the nearer thousandth, the number of square metres in 2 square feet.

5. Find, to the nearer hundredth, the number of square metres in 3 square yards.

6. A piece of copper bus bar is 0.75 millimetre by 100 millimetres in cross section. What is the cross-sectional area in square centimetres? _____

7. A square surface cover is 10.16 centimetres on each side. The knock-out in the center of the cover has a diameter of 1.27 centimetres. Find the area of the cover to the nearer hundredth centimetre. _____

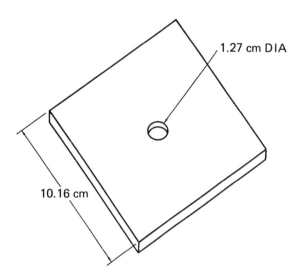

8. An electrician has to pour a concrete pad for a free standing Federal Pacific 1 600-ampere switch gear. The switch gear measures 75 centimetres wide, 240 centimetres long, and 230 centimetres high. If the pad has to be 30 centimetres larger than the switch on all sides, how many square metres of surface area will the pad cover? _____

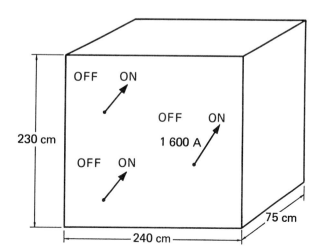

9. A show window requires 15 watts per square metre for sufficient lumination. How many 40-watt fluorescent fixtures are needed for the window shown? Round the answer to the nearer whole number. _____

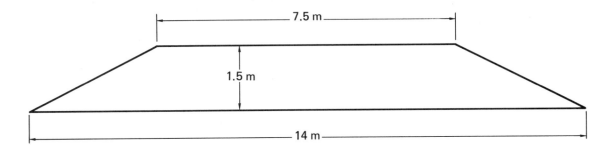

10. A rectangular concrete pad 180 centimetres wide and 450 centimetres long supports four transformers. Each transformer base measures 100 centimetres by 68 centimetres. What percent of the pad surface area is covered by the transformers? Round the answer to the nearer hundredth percent. _____

11. A switchboard 3 metres wide and 5 metres long is placed on a concrete slab near a wall. A live part extends 0.15 metre from the back of the switchboard. A clearance of 0.75 metre between the live part and the wall is required. There is a front walk space 1 metre wide and two side walkways each 0.6 metre wide. What is the area of the slab in square metres? _____

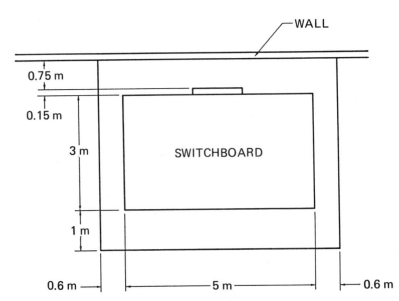

12. A type W4 conductor Neoprene portable cable has an outside diameter of 3.759 2 centimetres. What is the cross-sectional area of this cable? Express the answer to the nearer thousandth square centimetre. _____

13. The area of the floor in a building is 1 525 square metres. The floor is 25 metres wide. One strip of raceway is installed in the center length of the building. Find the number of metres of raceway that are needed. _____

14. An auditorium is 45 metres wide and 60 metres long. The building code requires 1.2 square metres for each person. What is the seating capacity of the auditorium? _____

15. A conduit has an inside diameter of 7.792 7 centimetres. What is the inside area? Round the answer to the nearer thousandth square centimetre. _____

16. Capacitance is directly proportional to the surface area of the plates. What is the total plate area, in square centimetres, of a capacitor with 6 plates of the same dimension shaped as shown? Round the answer to the nearer hundredth centimetre. _____

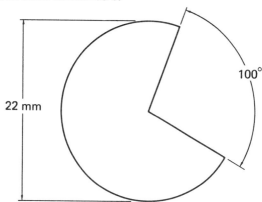

17. The power requirements to light a store area 60 feet by 20 feet is 5 watts per square foot.
    a. What is the total power requirement? a. _____
    b. The total load current is found by dividing the power by the voltage. Find the total load current, in amperes, needed for the area. The voltage is 120 volts. b. _____
    c. How many 15-ampere circuits are necessary to feed the area? c. _____

18. An imbedded heating cable is installed in a driveway to melt ice and snow. The driveway is 3 metres wide and 40.2 metres long. The cable uses 250 watts of power when activated. How many watts of power are used per square metre? Round the answer to the nearer hundredth. _____

Unit 27 VOLUME AND MASS MEASURE

BASIC PRINCIPLES OF VOLUME AND MASS MEASURE

Volume Measure

Volume is a three-dimensional measurement. The volume of an object can be found by multiplying its length, width, and height. The following charts show volume measurements for both the English and metric systems.

VOLUME MEASURE FOR SOLIDS

ENGLISH VOLUME MEASURE FOR SOLIDS
1 cubic yard (cu yd) = 27 cubic feet (cu ft)
1 cubic foot (cu ft) = 1 728 cubic inches (cu in)

METRIC VOLUME MEASURE FOR SOLIDS
1 000 cubic millimetres (mm^3) = 1 cubic centimetre (cm^3)
1 000 cubic centimetres (cm^3) = 1 cubic decimetre (dm^3)
1 000 cubic decimetres (dm^3) = 1 cubic metre (m^3)
1 000 cubic metres (m^3) = 1 cubic dekametre (dam^3)
1 000 cubic dekametres (dam^3) = 1 cubic hectometre (hm^3)
1 000 cubic hectometres (hm^3) = 1 cubic kilometre (km^3)

ENGLISH-METRIC VOLUME MEASURE FOR SOLIDS	
1 cubic inch (cu in)	= 16.387 064 cubic centimetres (cm^3)
1 cubic foot (cu ft)	≈ 0.028 317 cubic metre (m^3)
1 cubic yard (cu yd)	≈ 0.764 555 cubic metre (m^3)
1 cubic centimetre (cm^3)	≈ 0.061 024 cubic inch (cu in)
1 cubic metre (m^3)	≈ 35.314 667 cubic feet (cu ft)
1 cubic metre (m^3)	≈ 1.037 951 cubic yards (cu yd)

VOLUME MEASURE FOR FLUIDS

ENGLISH VOLUME MEASURE FOR FLUIDS
1 quart (qt) = 2 pints (pt)
1 gallon (gal) = 4 quarts (qt)

Unit 27 Volume and Mass Measure

METRIC VOLUME MEASURE FOR FLUIDS
10 millilitres (mL) = 1 centilitre (cL)
10 centilitres (cL) = 1 decilitre (dL)
10 decilitres (dL) = 1 litre (L)
10 litres (L) = 1 dekalitre (daL)
10 dekalitres (daL) = 1 hectolitre (hL)
10 hectolitres (hL) = 1 kilolitre (kL)

ENGLISH-METRIC VOLUME MEASURE FOR FLUIDS	
1 gallon (gal)	≈ 3 785.411 cubic centimetres (cm^3)
1 gallon (gal)	≈ 3.785 411 litres (L)
1 quart (qt)	≈ 0.946 353 litre (L)
1 ounce (oz)	≈ 29.573 530 cubic centimetres (cm^3)
1 cubic centimetre (cm^3)	≈ 0.000 264 gallon (gal)
1 litre (L)	≈ 0.264 172 gallon (gal)
1 litre (L)	≈ 1.056 688 quarts (qt)
1 cubic centimetre (cm^3)	≈ 0.033 814 ounce (oz)

SOLID-FLUID VOLUME EQUIVALENTS

ENGLISH VOLUME MEASURE EQUIVALENTS	
1 gallon (gal)	= 0.133 681 cubic foot (cu ft)
1 gallon (gal)	= 231 cubic inches (cu in)

METRIC VOLUME MEASURE EQUIVALENTS	
1 cubic decimetre (dm^3)	= 1 litre (L)
1 000 cubic centimetres (cm^3)	= 1 litre (L)
1 cubic centimetre (cm^3)	= 1 millilitre (mL)

Mass

The mass of an object indicates its density. Mass and weight are often confused as being the same thing. Mass and weight are proportional only when the object is in a gravitational field. A lead bar on earth could have a weight of 50 pounds, but in outer space it would have no weight. Its mass, however, would be unchanged.

MASS MEASURE

ENGLISH MASS MEASURE	
16 ounces (oz)	= 1 pound (lb)
2 000 pounds (lb)	= 1 ton

METRIC MASS MEASURE	
10 milligrams (mg)	= 1 centigram (cg)
10 centigrams (cg)	= 1 decigram (dg)
10 decigrams (dg)	= 1 gram (g)
10 grams (g)	= 1 dekagram (dag)
10 dekagrams (dag)	= 1 hectogram (hg)
10 hectograms (hg)	= 1 kilogram (kg)
1 000 kilograms (kg)	= 1 megagram (Mg)

ENGLISH-METRIC MASS MEASURE	
1 pound (lb)	≈ 0.453 592 kilogram (kg)
1 pound (lb)	≈ 453.592 37 grams (g)
1 ounce (oz)	≈ 28.349 523 grams (g)
1 ounce (oz)	≈ 0.028 350 kilogram (kg)
1 kilogram (kg)	≈ 2.204 623 pounds (lb)
1 gram (g)	≈ 0.002 205 pound (lb)
1 kilogram (kg)	≈ 35.273 962 ounces (oz)
1 gram (g)	≈ 0.035 274 ounce (oz)

PRACTICAL PROBLEMS

1. Find the number of cubic millimetres in 1.2 cubic metres.
2. Find the number of cubic metres in 2 400 cubic centimetres.
3. Find, to the nearer hundredth, the number of cubic yards in 42 cubic feet.
4. Find the number of cubic centimetres in 24 cubic millimetres.
5. Express 6 quarts to the nearer hundredth litre.
6. Express 3 gallons to the nearer thousandth litre.
7. Express 9 cubic yards to the nearer hundredth cubic metre.
8. Express 48 ounces to the nearer hundredth litre.
9. Express 5 cubic feet to the nearer thousandth cubic metre.
10. Express 7 cubic inches to the nearer hundredth cubic centimetre.
11. The diameter of an underground fuel tank is 2.6 metres and it is 7 metres long. How many gallons of fuel will the tank hold? Round the answer to the nearer gallon.

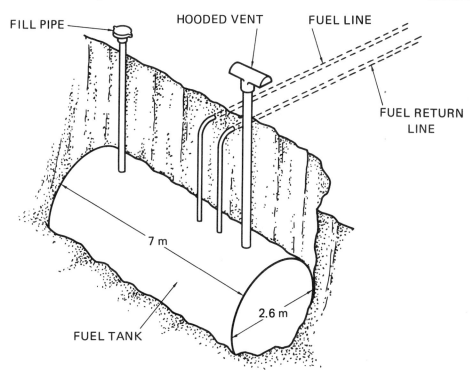

12. A power transformer with the capacity of 7 gallons of askarel is one-half full. How many additional litres of askarel are needed to fill the transformer? Round the answer to the nearer tenth. _____

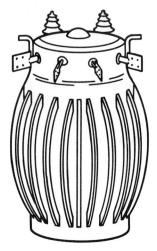

OIL-FILLED POWER TRANSFORMER

13. How many cubic metres of air are in a room 10 feet wide, 14 feet long, and 8 feet high? Round the answer to the nearer tenth. _____

Unit 28 ENERGY AND TEMPERATURE MEASURE

BASIC PRINCIPLES OF ENERGY AND TEMPERATURE MEASURE

Use the following charts to solve the problems presented in this unit.

MULTIPLICATION FACTOR	PREFIX	SYMBOL
1 000 000 = 10^6	mega	M
1 000 = 10^3	kilo	k
100 = 10^2	hecto	h
10 = 10	deka	da
0.1 = 10^{-1}	deci	d
0.01 = 10^{-2}	centi	c
0.001 = 10^{-3}	milli	m
0.000 001 = 10^{-6}	micro	μ
0.000 000 001 = 10^{-9}	nano	n
0.000 000 000 001 = 10^{-12}	pico	p

QUANTITY	UNIT	SYMBOL
power	kilowatt	kW
	watt	W
electric current	ampere	A
electromotive force	volt	V
electric resistance	ohm	Ω
energy	megajoule	MJ
	kilojoule	kJ
	joule	J
	kilowatt-hour (3.6 MJ)	kW·h
frequency	megahertz	MHz
	kilohertz	kHz
	hertz	Hz
electric capacitance	farad	F
inductance	henry	H

89

PRACTICAL PROBLEMS

Express the following as numbers between 1 and 10 times the proper power of 10:

1. 1.5 amperes
   a. _____ mA
   b. _____ µA

2. 12 volts
   a. _____ mV
   b. _____ µV

3. 2 microfarads
   a. _____ F
   b. _____ pF

4. 3 300 ohms
   a. _____ kΩ
   b. _____ MΩ

5. 37 henries
   a. _____ µH
   b. _____ mH

6. 0.33 kilowatt
   a. _____ W
   b. _____ MW

7. Express 68°F in degrees Celsius. _____

$$°C = \frac{5}{9}(°F - 32°)$$

8. A 20-ampere fuse element will open at 203°F due to heat from excessive current. At what Celsius temperature will the fuse open? _____

$$°C = \frac{5}{9}(°F - 32°)$$

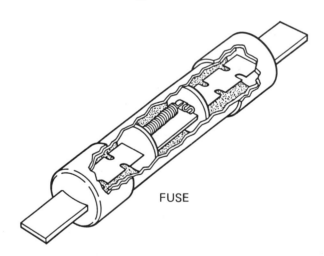

FUSE

9. A coffee percolator rated at 120 volts, 1 400 watts requires 15 minutes to complete its cycle. How many megajoules of work are used to make one pot of coffee? _____

10. The velocity or ratio wave propagation in free space is 186 000 miles per second. What is the velocity in kilometres per second? _____

11. What is the time required for one alternation of a frequency of 60 hertz? Round the answer to the nearer tenth millisecond. _____

$$t = \frac{1}{f}$$ where t = time in seconds
f = frequency in hertz
1 = constant one second

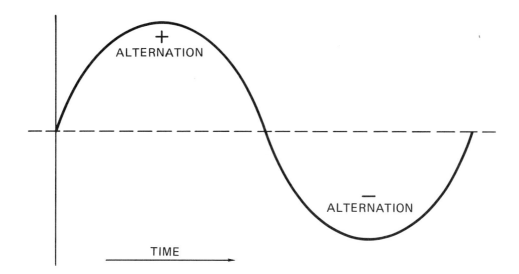

12. The current flow past a given point is 1.758 x 10^{19} electrons per second. Find, to the nearer tenth ampere, the current flow past this point. _____

    1 coulomb = 6.28 x 10^{18} electrons
    1 ampere = 1 coulomb/sec

13. Calculate the wavelength, in metres, for a frequency of 60 hertz. _____

$$\lambda = \frac{3 \times 10^8}{f}$$ where λ = wavelength
f = frequency in hertz
3×10^8 = a mathematical constant

14. The measured inductance of a power line 110 kilometres long is found to be 295 millihenries.

    a. What is the inductance of the line per kilometre? Express the answer to the nearer hundredth millihenry.   a. _____

    b. What is the inductance of the line per metre? Express the answer to the nearer hundredth microhenry.   b. _____

Unit 29 COMBINED PROBLEMS ON MEASURE

BASIC PRINCIPLES OF COMBINED PROBLEMS ON MEASURE

This unit provides practical problems involving combined problems on measure.

PRACTICAL PROBLEMS

Express each of the following in the indicated units. Round the answer to the nearer thousandth when necessary.

1. 12 inches as millimetres. _____
2. 8 yards as metres. _____
3. 5 miles as kilometres. _____
4. 10 pints as litres. _____
5. 6 gallons as litres. _____
6. 20 square feet as square metres. _____
7. 15 cubic yards as cubic metres. _____
8. 80 pounds as kilograms. _____
9. 68 degrees Fahrenheit as degrees Celsius. _____
10. 0.740 kilowatt-hour as kilojoules. _____
11. 10 kilowatt-hours as megajoules. _____
12. 500 000 milliamperes as amperes. _____
13. 200 kilovolts as megavolts. _____
14. 5 000 microfarads as farads. _____
15. A drum has a radius of 24 inches and a length of 3 1/2 feet. How many litres does the drum hold? _____
16. The diameter of #10 AWG wire is 0.028 5 inch. What is the radius in centimetres? _____
17. The bed of a dump truck measures 5 feet deep, 5 feet wide, and 10 feet long. When filled level, how many cubic metres of fill does the truck hold? _____
18. A power transformer needs 25 gallons of askarel. Find the number of one-litre containers of askarel needed for the transformer. _____

19. The inside diameter of a piece of tubing is 5.5 centimetres and the outside diameter is 6.3 centimetres. What is the thickness of the walls of the tubing? _____

20. It costs a contractor $90 to drive his pick-up truck 500 kilometres. Find the cost per kilometre to operate the truck. _____

21. A bushing 3 centimetres long is cut to a length of 2.25 centimetres. How many centimetres are cut off? _____

22. An office 7 metres by 6 metres is wired for lights. Twenty 40-watt lamps are installed. What is the power consumption for the office lighting in watts per square metre? _____

23. A building 50 metres long occupies a surface area of 1 250 square metres. Find the width of the building. _____

24. A piece of copper bus bar is 1 centimetre thick by 8 centimetres wide. Find the cross-sectional area. _____

25. A welding cable has an outside diameter of 1.1 millimetres. Find the cross-sectional area of this cable. _____

26. A box cover is 10.16 centimetres in diameter and has a 20.42-millimetre hole in the center. Find the area of the cover. _____

Ratio and Proportion

SECTION 7

 Unit 30 RATIO

BASIC PRINCIPLES OF RATIO

Ratio

Ratio is the comparison of two numbers or quantities as a quotient. Ratios, like fractions, are generally reduced to lowest terms. The ratio 16:8 (read as 16 to 8) can be reduced to a ratio of 2:1 by dividing both numbers by 8. Ratios are also written in fractional form. The ratio 5:9 can also be written as

$$\frac{5}{9}$$

Inverse Ratios

An *inverse ratio* is the ratio in reverse order of the original ratio. The inverse ratio of 5:9 is 9:5 or

$$\frac{9}{5}$$

PRACTICAL PROBLEMS

1. Express each ratio in lowest terms.
   a. 15:5
   b. 25:10
   c. 12:4
   d. 3/4:1/4
   e. 25:75

   a. _____
   b. _____
   c. _____
   d. _____
   e. _____

2. Find the inverse of each ratio.
   a. 10:3
   b. 5:2
   c. 7:8
   d. 7/8:1/8
   e. 5:12

   a. _____
   b. _____
   c. _____
   d. _____
   e. _____

3. What is the ratio of the number of primary turns to the number of secondary turns in the following diagram? _____

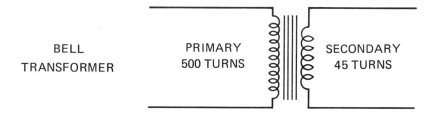

4. What is the ratio of the speed of one motor turning at 1 750 revolutions per minute to the speed of a second motor turning at 3 500 revolutions per minute? _____

5. What is the ratio of one generator with an output of 3 500 watts to a second generator with an output of 24 500 watts? _____

6. If it takes one electrician 18 hours to wire a house and a second electrician 45 hours to wire a similar house, what is the ratio of the second electrician's time to the first electrician's time? _____

7. What is the ratio of a pinion gear with 14 teeth to a driven gear with 72 teeth? _____

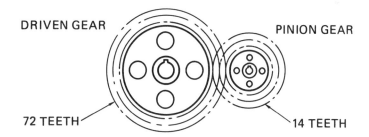

Unit 31 PROPORTION

BASIC PRINCIPLES OF PROPORTION

Proportion is the equality of two ratios.

$$5:2 = 25:10$$

There are two basic methods of solving problems dealing with proportion. One method is to multiply the means together and the extremes together. The means are the two inside terms and the extremes are the two outside terms.

$$5:25 = 3:x$$

with means being 25 and 3 (inside), and extremes being 5 and x (outside).

To solve this problem, multiply the means together and the extremes together.

$$5x = 75$$

To find the value of x, divide both sides by 5.

$$x = 15$$

Another method of solving this problem is to use cross multiplication. When cross multiplication is used, the two ratios are written as fractions separated by an equal sign.

$$\frac{5}{25} = \frac{3}{x}$$

The top part of one ratio is then multiplied by the bottom part of the other.

$$\frac{5}{25} \diagup\!\!\!\diagdown \frac{3}{x}$$

$$5x = 75$$

$$x = 15$$

PRACTICAL PROBLEMS

1. A motor-driven pump discharges 306 gallons of water in 3.6 minutes. How long will it take to discharge 5 200 gallons? Express the answer to the nearer tenth. _____

2. If a piece of cable 160 feet long costs $60.00, what will 500 feet cost at the same rate? _____

3. An electrically-driven sump pump discharges 125 gallons of water in 4.5 minutes. What time will it take to discharge 320 gallons? _____

4. A copper wire 750 feet long has a resistance of 1.893 ohms. How long is a copper wire of the same area whose resistance is 3.156 ohms? Express the answer to the nearer tenth. _____

5. A wire whose resistance is 5.075 ohms has a diameter of 31.961 mils. What is the resistance of a wire of the same material and length, if the diameter is 40.303 mils? Resistance varies inversely as the square of the diameter. Round the answer to the nearer thousandth. _____

$$\frac{R_1}{R_2} = \frac{d_2^2}{d_1^2}$$

6. A wire 2 725 feet long, 85 mils in diameter, has a resistance of 0.372 ohm. Find, to the nearer thousandth, the resistance of 3 600 feet of the same wire. _____

7. If a wire 1 325 feet long has a resistance of 0.65 ohm, what is the resistance, to the nearer hundredth, of one mile of the same wire? _____

8. Four workers complete a certain electrical job in 120 hours. How long will it take three workers to do the same type of job? _____

9. One hundred twenty feet of conduit cost $15.89. What will 250 feet of conduit cost at the same rate? _____

10. An electrical repair team of 12 workers complete a job in 288 hours. How long should it take 9 workers to do the same amount of work? _____

11. If 120 feet of 2-inch conduit cost $80.50, what will 325 feet of 2-inch conduit cost? _____

Unit 32 COMBINED OPERATIONS WITH RATIO AND PROPORTION

BASIC PRINCIPLES OF COMBINED OPERATIONS WITH RATIO AND PROPORTION

This unit provides practical problems involving combined operations with ratio and proportion.

PRACTICAL PROBLEMS

1. A piece of cable 8 feet long costs $60.00. What will 10 feet cost at the same rate?

2. A wire, 100 feet long, has a resistance of 800 ohms. How long is a copper wire of the same area whose resistance is 600 ohms?

3. Conduit costs $1 000 for 500 feet. Find the cost of 200 feet at the same rate.

4. An electrical repair team of 6 electricians complete a job in 3 hours. How long should it take 2 electricians to do the same amount of work?

5. One motor has a speed of 1 800 revolutions per minute. A second motor turns at 3 600 revolutions per minute. What is the ratio of the speed of the first motor to that of the second motor?

6. It costs $120 for 10 feet of 4-inch conduit. What will it cost for 50 feet of 4-inch conduit?

Note: Use this diagram for problems 7-8.

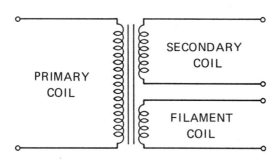

7. Find the ratio of 10 turns of the filament coil to 200 turns of the primary coil.

8. Find the ratio of 200 turns of the primary coil to 800 turns of the secondary coil.

9. An electric drill has a chuck attachment to obtain a slower speed. If the attachment has a ratio of 4:1, find the resulting speed if the drill normally turns at 3 600 revolutions per minute.

10. A new "dry pump" is capable of delivering 100 gallons of liquid per minute. How long will it take to deliver 6 000 gallons?

Formulas

SECTION 8

Unit 33 REPRESENTATION IN FORMULAS

BASIC PRINCIPLES OF REPRESENTATION IN FORMULAS

A *formula* is a mathematical statement of equality. Just as words are written using symbols, formulas are also written using symbols. In the following problems, written statements will be expressed as mathematical formulas, and formulas will be expressed as written statements.

When writing a formula, it is helpful to know what some of the statements mean and how some of the symbols are used to indicate different operations. The word *sum* means to add, and a plus sign (+) is used to indicate addition.

The word *product* is used to indicate multiplication. There are several methods used to indicate that two quantities are to be multiplied together. One method is to simply write two letters together with no sign between them. IR means to multiply I times R. Another method is to use parentheses around the values to be multiplied together. $(I)(R)$ means to multiply I and R together. Symbols are also used to represent multiplication. The multiplication sign (×) is often used between numbers, but is seldom used with alphabetical characters. The multiplication dot (·) is often used between two alphabetical characters to represent multiplication. Another symbol used to represent multiplication, the asterisk (*), has become popular because of computers.

The words ratio and *inversely proportional* are represented by division. Division is generally indicated in a formula by writing the numbers as a fraction. If the letter E is to be divided by R, it would be written as

$$\frac{E}{R}$$

Subtraction is generally indicated by statements such as *the difference of*. The minus sign (−) is generally used to represent subtraction.

Another term often used in mathematics is *reciprocal*. The term reciprocal means 1 divided by the number. The reciprocal of 4 is

$$\frac{1}{4}$$

PRACTICAL PROBLEMS

Write a mathematical formula for each statement. Use the symbols given in the problem to write each formula.

1. The total resistance (R_t) of a series electrical circuit is equal to the sum of the individual resistances ($R_1, R_2, \ldots, R_n$). _____

2. The total capacitance (C) is directly proportional to the product of the dielectric constant (k) and plate area (A) and inversely proportional to the thickness (d) of the dielectric. _____

3. In a power transformer the ratio of the primary voltage (E_p) to the secondary voltage (E_s) is equal to the ratio of the power in the primary (P_p) to the power in the secondary (P_s). _____

4. The current (I) in an electrical circuit is directly proportional to the voltage applied (E_a) and inversely proportional to the circuit impedance (Z). _____

5. The capacitance (C) may be found by taking the reciprocal of the product of two pi (π), the frequency (f), and the capacitive reactance (X_c). _____

6. An electrical motor has a degree of efficiency (Eff) equal to the ratio of its useful output power (P_o) to the input power (P_i) required to operate it. _____

7. The electrical power (P) dissipated by an incandescent lamp is directly proportional to the square of the voltage drop (E) across the lamp and inversely proportional to the electrical resistance (R) of the lamp. _____

8. The electrostatic force (F) between two plates of a charged capacitor is directly proportional to the product of the charges (Q_1 and Q_2) on each plate and inversely proportional to the square of the distance (d) between the plates. _____

9. Between two coils there exists a mutual inductance (M) that is equal to the square root of the product of the two inductances (L_1 and L_2) multiplied by the coefficient of coupling (k) between the two coils. _____

10. The resonant frequency (f_r) of a circuit containing both inductance (L) and capacitance (C) may be found by the reciprocal of two Pi (π) multiplied by the square root of the product of the inductance and capacitance. _____

Write a statement for each of the formulas given.

11. $G = \dfrac{1}{R}$ where G = conductance
R = resistance

___

12. $R_s = \dfrac{I_m R_m}{I_s}$ where R_s = resistance of the ammeter shunt
I_m = current through the meter movement
R_m = resistance of the meter movement
I_s = current through the shunt

___

13. $E = 0.707 \times E_{max}$ where E = effective value of AC voltage
E_{max} = maximum peak value of AC voltage
0.707 = a derived constant

___

14. $PF = \dfrac{R_t}{Z}$ where PF = power factor of an AC circuit
R_t = total circuit resistance
Z = circuit impedance

___

15. $f = \dfrac{PN}{60}$ where f = frequency of the generated AC voltage, in hertz
P = number of pairs of poles of the generator
N = revolutions per minute of the generator in the magnetic field
60 = a constant representing the number of seconds in a minute

Unit 34 REARRANGEMENT IN FORMULAS

BASIC PRINCIPLES OF REARRANGING FORMULAS

A mathematical formula is generally referred to as an *equation*. The word equation means equal to or to equate. An equation is a statement of equality which indicates that everything on one side of the equal sign is equal to everything on the other side of the equal sign. A pair of balance scales is a good example of this principle. As long as the weight is the same on each side of the scales, they will be in balance. Weight can be added to or removed from the scales without affecting the balance as long as the same amount of weight is added to or removed from both sides.

The same principle is true for an equation. Any operation can be performed on one side of the equation as long as the same operation is performed on the other side. In the following problems, formulas will be rearranged to find different quantities. This is done by rearranging the formula so that the quantity to be found is on one side of the equal sign by itself. When rearranging formulas, remember two principles.

1. Anything can be done to one side of an equation as long as the same thing is done to the other side.

2. When removing factors from one side of the equation perform the opposite function that is indicated.

Example: If $X = 2JLC$. Solve for L.

To solve for L, it must be on one side of the equal sign by itself. This can be done by removing all the other factors on the right side of the equation. The formula states that X is equal to 2 times J times L times C. Multiplication is the function indicated. To remove all the factors, except L, perform the opposite function which is division.

$$\frac{X}{2JC} = \frac{2JLC}{2JC}$$

Notice that both sides of the equation are divided by $2JC$.

$$\frac{X}{2JC} = \frac{\cancel{2JC}\,L}{\cancel{2JC}\,1}$$

The equation is now

$$\frac{X}{2JC} = \frac{L}{1}$$

The formula can now be written:

$$L = \frac{X}{2JC}$$

PRACTICAL PROBLEMS

Solve each of the equations for the variable shown.

1. $Q = CV$, solve for C.

2. $I = \frac{E}{Z}$, solve for Z.

3. $R^2 = Z^2 - X^2$, solve for Z.

4. $R = \frac{P}{I^2}$, solve for I.

5. $R_2 = R_t - R_1 - R_3$, solve for R_t.

6. $X_L = 2\pi fL$, solve for L.

7. $X_C = \frac{1}{2\pi fC}$, solve for C.

8. $pf = \frac{R}{X}$, solve for R.

9. $N_s = \frac{E_s N_p}{E_p}$, solve for N_p.

10. $P = \frac{120f}{N}$, solve for f.

11. $\frac{Z_p}{Z_s} = \frac{N_p^2}{N_s}$, solve for Z_p.

12. $E_s I_s = E_p I_p$, solve for E_p.

13. $u = g_m r_p$, solve for r_p.

14. $C = \frac{0.088\ 4kA(N-1)}{d}$, solve for A.

15. $M = k\sqrt{L_1 L_2}$, solve for k.

Unit 35 GENERAL SIMPLE FORMULAS

BASIC PRINCIPLES OF SIMPLE FORMULAS

To solve a formula, the alphabetical characters are replaced with numbers from known values. Multiplication, division, addition, and subtraction are then performed as indicated by the formula.

Example: An electric iron has a current draw of 5 amperes and a resistance of 10 ohms. How much power is being consumed by this iron? The formula for this problem is $P = I^2 R$ where

$$P = \text{power}$$
$$I = \text{current}$$
$$R = \text{resistance}$$

To solve the problem, substitute number values for the letter values in the formula.

$$P = 5 \times 5 \times 10$$
$$P = 250 \text{ watts}$$

PRACTICAL PROBLEMS

1. Find the total resistance (R_t) of the three field rheostats shown. Express the answer to the nearer hundredth.

$$R_t = \frac{1}{\frac{1}{R_1} + \frac{1}{R_2} + \frac{1}{R_3}}$$

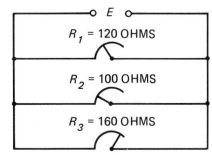

2. A motor takes 38 amperes on a 220-volt circuit. Find the horsepower output of the motor shown with an efficiency of 90%. Express the answer to the nearer hundredth.

$$I = \frac{hp \times 746}{E \times Eff}$$

where hp = horsepower
E = volts
I = amperes
Eff = efficiency

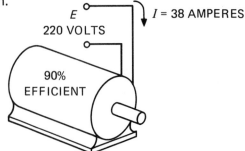

104

3. A circuit uses 124 amperes at a voltage of 230 volts. Find the load in kilowatts.

$$P = \frac{E \times I}{1\,000}$$

where P = kilowatts
E = volts
I = amperes

4. Find the resistance of 240 feet of number 10 copper wire with a cross-sectional area of 10 380 circular mils. Express the answer to the nearer hundredth.

$$R = \frac{K \times L}{d^2}$$

where R = resistance
K = specific resistance of copper (10.8)
L = length in feet
d = diameter in mils

5. A cell has a voltage of 2.4 volts and an internal resistance of 0.03 ohm. The cell is connected to an electromagnet with a resistance of 1.5 ohms. What current, to the nearer hundredth, does the magnet receive?

$$I = \frac{E}{R + r}$$

where I = amperes
E = volts
r = internal resistance
R = resistance of external circuit

6. What is the resistance of R_1 if the total resistance (R_t) of the parallel circuit shown is 3.2 ohms?

$$R_t = \frac{1}{\frac{1}{R_1} + \frac{1}{R_2} + \frac{1}{R_3} + \frac{1}{R_4}}$$

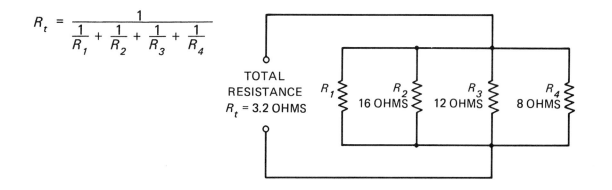

TOTAL RESISTANCE
R_t = 3.2 OHMS

R_1 R_2 16 OHMS R_3 12 OHMS R_4 8 OHMS

7. The impressed voltage across a motor armature is 230 volts and the armature resistance is 0.48 ohm. The counter voltage equals 224 volts. What current does the armature take?   _____

$$I = \frac{E_x - E_c}{R}$$

where I = amperes
R = resistance
E_x = impressed voltage
E_c = counter voltage

8. The internal resistance of five cells is 1.8 ohms each. The cells are connected as shown in the diagram to a 2-ohm resistance. What current, to the nearer hundredth, exists in the circuit?   _____

$$I = \frac{E \times ns}{(r \times ns) + R}$$

where E = volts of one cell
ns = number of cells in series
r = internal resistance of one cell
R = resistance of external circuit
I = amperes

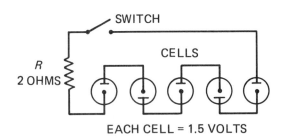

EACH CELL = 1.5 VOLTS

Note: Use this formula for problems 9-11.

$$°F = \frac{9}{5}(°C) + 32°$$

where $°C$ = temperature in degrees Celsius
$°F$ = temperature in degrees Fahrenheit

9. The temperature of an oven is 200 °C. Find the temperature of the oven in degrees Fahrenheit.   _____

10. The temperature of a motor rises to 50 °C above room temperature. The room temperature is 26 °C. What will a Fahrenheit thermometer read?   _____

11. If the temperature of a commutator is 65 °C, what is the temperature in degrees Fahrenheit?   _____

12. If R_x is the unknown resistance of the Wheatstone bridge arrangement shown in the figure, find the value of this unknown resistance.   _____

$$\frac{R_x}{R_3} = \frac{R_1}{R_2}$$

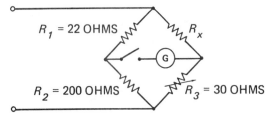

13. The voltage drop across a 100-ohm standard resistance is 16 volts. What is the resistance of a coil connected in series if the voltage drop across the coil is 8 volts?

$$\frac{V_R}{V_x} = \frac{R}{R_x}$$ where V_R = voltage drop across standard resistance
V_x = voltage drop across unknown resistance
R = resistance of standard in ohms
R_x = resistance of coil in ohms

14. Determine the horsepower output, to the nearer hundredth, of the motor in the circuit shown.

$$\text{hp input} = \frac{EI - 2R(I)^2}{746}$$ where E = voltage in volts
I = current in amperes
R = resistance in ohms

hp output = hp input × efficiency

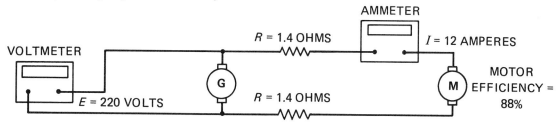

15. A motor turns 1 130 revolutions per minute (r/min) with a 6-inch driving pulley (d). What are the revolutions per minute of a 24-inch pulley (D) belted to this motor?

r/min of D = $\dfrac{\text{Diameter of } d \times \text{r/min of } d}{\text{Diameter of } D}$

16. Find the power used in the single phase circuit shown.

$P = E \times I \times$ power factor (PF)

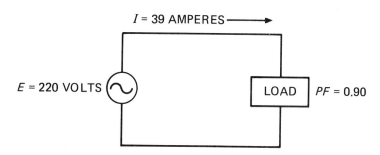

17. A 2-phase, 4-wire, 220-volt circuit has a current of 16.5 amperes per leg. The power factor is 0.85. Find the number of watts consumed. _____

For a 2-phase, 4-wire circuit:
$$P \text{ (watts)} = 2 \times E \times I \times PF$$

18. A motor takes a current of 27.5 amperes per leg on a 440-volt, 3-phase circuit. The power factor is 0.80. What is the load in watts? Round the answer to the nearer whole watt. _____

For a 3-phase circuit:
$$P \text{ (watts)} = \sqrt{3} \times E \times I \times PF$$

19. What is the current per leg in a 2-phase, 3-wire circuit if the power is 200 watts, the voltage is 120 volts, and the PF is 0.9? Express the answer to the nearer thousandth. _____

$$P \text{ (watts)} = 2 \times E \times I \times PF$$

Note: Use this formula for problems 20-22.

$$hp = \frac{I \times E \times Eff}{746}$$ where I = current in amperes
E = voltage in volts
Eff = efficiency of motor
hp = horsepower

20. Solve the equation for the current required by a motor when the horsepower, percent efficiency of the motor, and the voltage are known. _____

21. A 12-horsepower motor is connected to a 220-volt circuit. How much current, to the nearer hundredth, does it take if the motor efficiency is 87%? _____

22. If a 25-horsepower motor takes 96 amperes at full load, and the motor efficiency is 90%, what is the terminal voltage? Express the answer to the nearer hundredth. _____

Note: Use this formula for problems 23-24.

$$Z = \sqrt{R^2 + X^2}$$ where Z = impedance in ohms
R = resistance
X = reactance

Unit 35 General Simple Formulas 109

23. What is the impedance, to the nearer hundredth, of the circuit shown in the figure?

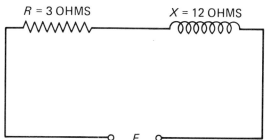

24. Find the impedance of a coil with 5.5 ohms resistance and 8.2 ohms reactance? Round the answer to the nearer hundredth.

Note: Use this formula for problems 25-26.

$$Z = \frac{E}{I}$$

where Z = impedance in ohms
E = volts
I = amperes

25. A circuit has a voltage of 440 volts across it and a current of 5.5 amperes. What is the impedance of the circuit?

26. If a current of 4.5 amperes exists, what is the impedance of a circuit having a voltage of 110 volts across it? Express the answer to the nearer hundredth.

Note: Use this formula for problems 27-28.

$$kW \text{ load} = \frac{I \times E \times \sqrt{3} \times PF}{1\,000}$$

where kW = 3-phase power in kilowatts
I = current in amperes
E = voltage in volts
PF = power factor

27. A 3-phase, 3-wire circuit, supplying power to several motors, delivers 275 amperes. If the voltage is 220 volts, and the power factor is 0.80, what is the kilowatt load? Round the answer to the nearer hundredth.

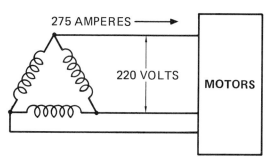

28. A 3-phase, 3-wire circuit delivers 175 amperes at a voltage of 220 volts with a 0.85 power factor. What is the kilowatt load to the nearer hundredth? _____

29. A 2-phase, 4-wire circuit with voltage of 220 volts is connected to a group of motors that requires 85 amperes with a 0.80 power factor. What is the load in kilowatts? _____

$$kW \text{ load} = \frac{I \times E \times 2 \times PF}{1\,000}$$

where kW = 2-phase power in kilowatts
I = amperes
E = volts
PF = power factor

30. A 3-wire, 2-phase circuit has a voltage of 220 volts, current of 85 amperes and a power factor of 0.80.

   a. What is the load in kilowatts to the nearer hundredth? a. _____

$$kW \text{ load} = \frac{I \times E \times \sqrt{2} \times PF}{1\,000}$$

where kW = 2-phase power in kilowatts
I = amperes
E = volts
PF = power factor

   b. What is the horsepower output if the overall efficiency is 82%? Round the answer to the nearer hundredth. b. _____

$$hp = \frac{I \times E \times 2 \times \text{Eff} \times PF}{746}$$

where I = line current
E = phase voltage
hp = horsepower
Eff = efficiency
PF = power factor

31. What is the current in amperes in a 3-phase circuit, when the load is 67.5 horsepower, the efficiency is 90%, the voltage is 440 volts, and the circuit power factor is 0.80? Round the answer to the nearer hundredth. _____

$$hp = \frac{I \times E \times \sqrt{3} \times \text{Eff} \times PF}{746}$$

32. A 2-wire circuit supplying power to a load 180 feet from the meter has a 6.5 voltage drop. If the load is 37.5 amperes, what, to the nearer hundredth, is the circular mil area (CM) of the wire? _____

$$CM = \frac{K \times N \times L \times I}{E_e}$$

where CM = circular mil area of conductor
  L = length of one wire in feet
  K = constant for copper (10.8)
  I = amperes
  E_e = voltage drop
  N = number of wires

33. In a circuit, what current does a 220-volt, 5 horsepower motor receive if the efficiency of the motor is 80%? Express the answer to the nearer hundredth. _____

$$hp = \frac{I \times E \times Eff}{746}$$

where hp = horsepower
  I = amperes
  E = volts
  Eff = efficiency of motor

34. In a 2-wire circuit what size conductors, to the nearer hundredth, does a 220-volt, 5 horsepower motor require if the voltage drop is 3 volts and the distance is 160 feet from the meter? The motor has an 80% efficiency. _____

$$CM = \frac{hp \times 746 \times N \times L \times K}{E_e \times Eff \times E_m}$$

where CM = circular mil area of conductor
  hp = horsepower
  L = length of line wires in feet
  E_e = voltage drop
  K = constant for copper (10.8)
  Eff = efficiency of motor
  E_m = motor voltage
  N = number of wires

112 Section 8 Formulas

35. What size conductor should be used in a 2-wire system to service a load of 6 500 watts, 150 feet from the generator? The voltage at the load is 115 volts, and the voltage drop for one conductor is 2.5 volts. Express the answer to the nearer hundredth. _____

$$I = \frac{P}{E}$$

$$CM = \frac{K \times N \times L \times I}{E_e}$$

where I = amperes
P = watts
E = load voltage
CM = circular mil area of conductor
L = length of one wire in feet
K = constant for copper (10.8)
E_e = voltage drop
N = number of wires

36. Find the total voltage (E) for the circuit shown. Express the answer to the nearer tenth. _____

$$E = \sqrt{E_R^2 + (E_L - E_C)^2}$$

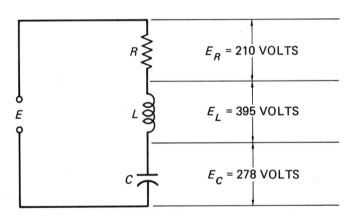

E_R = 210 VOLTS
E_L = 395 VOLTS
E_C = 278 VOLTS

37. What is the frequency in hertz (Hz) of the current furnished by an alternator having 8 poles and running at a speed of 900 revolutions per minute? _____

$$f = P \times \frac{N}{60}$$

where f = frequency in hertz
P = number of pairs of poles
N = revolutions per minute

Unit 36 OHM'S LAW FORMULAS

BASIC PRINCIPLES OF OHM'S LAW FORMULAS

To solve problems using Ohm's law substitute numerical values for alphabetical values in the formulas. It will be necessary to choose the proper formula to be used. This is done by determining what quantity is to be found, and what quantities are known.

Example: An electric toaster has a resistance of 10 ohms and draws a current of 12 amperes when connected to the line. What voltage is the toaster connected to?

The value to be found is voltage (E). The values known are resistance (R) and current (I). The formula chosen must contain these three quantities. The formula $E = IR$ will be used to solve the problem where

$$E = \text{voltage}$$
$$I = \text{current}$$
$$R = \text{resistance}$$

$$E = IR$$
$$E = 12 \times 10$$
$$E = 120 \text{ volts}$$

This formula can be rearranged as follows.

$$E = IR \qquad \text{where} \quad E = \text{volts}$$
$$R = \frac{E}{I} \qquad\qquad\quad R = \text{ohms}$$
$$I = \frac{E}{R} \qquad\qquad\quad I = \text{amperes}$$

PRACTICAL PROBLEMS

Express the answer to the nearer hundredth, when necessary.

1. How much current (I) flows through a lamp that has a resistance (R) of 220 ohms and is connected across a 115 volt (E) circuit? _____

2. What voltage is required to force 4.5 amperes through an electric iron having a resistance of 25.5 ohms? _____

114 Section 8 Formulas

Note: Use this diagram for problems 3-4.

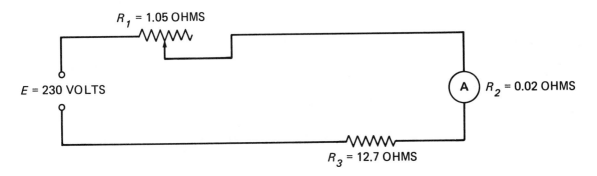

3. What current flows through the circuit shown in the diagram, if the total resistance in ohms equals $R_1 + R_2 + R_3$? _____

4. Find the value of current for the circuit, if $R_1 = 0$, and the total resistance in ohms is equal to $R_1 + R_2 + R_3$ _____

5. An electric heater radiates sufficient heat at a 10-ampere load on a 110-volt circuit. What is the hot resistance? _____

6. A lamp with a resistance (hot) of 50 ohms is connected across 115 volts. What current does the lamp receive? _____

7. The 2.4-volt cell with an internal resistance of 0.5 ohm is connected to an external resistance of 1.25 ohms. What current does the external resistance, R_L, receive? _____

$R = r + R_L$ where r = internal resistance
                        R_L = external resistance

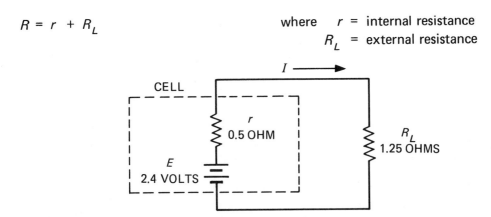

8. A lamp requires a current of 0.94 ampere when connected to a circuit with a voltage of 115 volts. Find the resistance, in ohms, of the lamp. _____

Note: Use this diagram for problems 9-10.

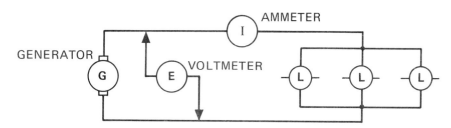

9. In the circuit shown, three lamps have a total resistance of 220 ohms. The voltmeter reading (E) is 115 volts. Find the ammeter reading (I) in this circuit. _____

10. Find the total resistance, in ohms, if the current in the circuit is 2.82 amperes, and the voltage is 115 volts. _____

11. The total resistance of the two-line wires is 0.314 ohm. The cross-sectional area of the wire is 16 510 circular mils, which equals d^2. What is the length of each wire? _____

$$R = \frac{K \times L}{d^2}$$

where R = resistance in ohms
K = 10.8 (constant for copper)
L = length of wire in feet
d = diameter in mils

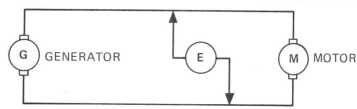

12. If the cross-sectional area of a wire is 16 510 circular mils, what is the diameter d of the wire? _____

Cross-sectional area = d^2 where d = diameter in mils

Note: Use this formula for problems 13-14.

$$R = \frac{K \times L}{d^2}$$

where R = resistance in ohms
L = length in feet
d = diameter in mils
K = constant, 10.8

13. What is the length of a wire having a resistance of 1.085 ohms and a diameter of 162.02 mils? _____

14. Find the length of a copper wire 0.104 inch in diameter, if it has a resistance of 3.1 ohms? Use 1 mil = 0.001 inch.

15. What is the coil resistance of an electric bell, if 0.25 ampere exists when a voltage of 2.8 volts is applied?

16. The four dry cells are connected in series as shown.

$E_T = e_1 + e_2 + e_3 + e_4$
$R_T = r_1 + r_2 + r_3 + r_4$

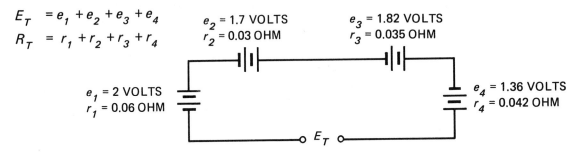

$e_1 = 2$ VOLTS
$r_1 = 0.06$ OHM

$e_2 = 1.7$ VOLTS
$r_2 = 0.03$ OHM

$e_3 = 1.82$ VOLTS
$r_3 = 0.035$ OHM

$e_4 = 1.36$ VOLTS
$r_4 = 0.042$ OHM

a. Find the total resistance.

b. Find the total voltage.

a. _____

b. _____

17. Using Ohm's Law, find the resistance of one conductor of an annunciator cable if the voltage across the conductor is 1.2 volts, and the current is 2 amperes.

18. A 60-volt electric time clock circuit with several clocks connected in parallel has a 2-wire line voltage drop of 12.5 volts when the circuit is closed. Using Ohm's Law, find the total resistance of the 2-wire line if a current of 10 amperes exists.

19. What is the resistance of an electromagnet connected to a cell having a voltage of 2.2 volts and drawing a current of 0.3 ampere?

20. Find the total resistance (R_T) of the circuit shown.

$R_T = R_1 + \dfrac{R_2 \times R_3}{R_2 + R_3}$

$R_1 = 58$ OHMS
$R_2 = 20$ OHMS
$R_3 = 30$ OHMS

Unit 36 Ohm's Law Formulas 117

21. Eight resistors are connected as shown. The value of each resistor is 0.5 ohm. Find the total resistance, R_T. _____

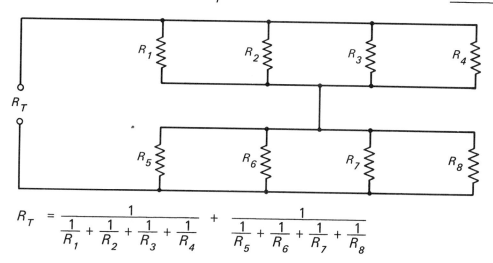

$$R_T = \frac{1}{\frac{1}{R_1} + \frac{1}{R_2} + \frac{1}{R_3} + \frac{1}{R_4}} + \frac{1}{\frac{1}{R_5} + \frac{1}{R_6} + \frac{1}{R_7} + \frac{1}{R_8}}$$

22. Find the total resistance (R_T) of the three low voltage relay coils. Round the answer to the nearer hundredth. _____

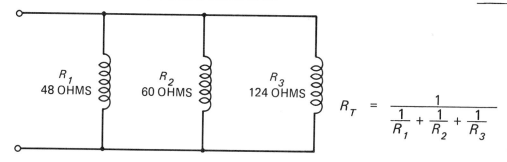

$$R_T = \frac{1}{\frac{1}{R_1} + \frac{1}{R_2} + \frac{1}{R_3}}$$

23. A wire has an area of 10 380 circular mils. Find the resistance of 900 feet of wire supplying charging current for a 24-volt battery. Express the answer to the nearer thousandth. _____

$$R = \frac{K \times L}{A}$$

where A = circular mils
K = 10.8
L = total length of wire in feet
R = resistance in ohms

24. One 2-wire section of a 24-volt watchman clock signal system is operated through a 450-foot length of lead-covered cable. The size of the conductor is number 18 and the diameter is 40.3 mils. What is the resistance of one conductor? Express the answer to the nearer thousandth. _____

$$R = \frac{N \times K \times L}{d^2}$$
where N = number of wires
K = constant (10.8)
L = length of one wire in feet
R = resistance in ohms
d = diameter in mils

25. A 30-conductor control cable 2 150 feet long is installed between the engine room and a small substation. The voltage drop is 5.5 volts and the line current is 12 amperes for one 2-wire circuit during operation. What is the size of the wire in circular mils (CM) for one 2-wire circuit? _____

$$CM = \frac{N \times K \times I \times L}{E_e}$$
where N = number of wires
E_e = voltage drop
L = length in feet per wire
CM = area in circular mils
I = current in amperes
K = 10.8

Unit 37 POWER FORMULAS

BASIC PRINCIPLES OF POWER FORMULAS

The following problems will be solved using the power formulas.

$$P = EI \qquad I = \frac{E}{R} \qquad E = IR \qquad R = \frac{E}{I}$$

$$P = I^2 R \qquad I = \frac{P}{E} \qquad E = \frac{P}{I} \qquad R = \frac{E^2}{P}$$

$$P = \frac{E^2}{R} \qquad I = \sqrt{\frac{P}{R}} \qquad E = \sqrt{PR} \qquad R = \frac{P}{I^2}$$

PRACTICAL PROBLEMS

1. A circuit is connected to a voltage (E) of 230 volts and has a current (I) of 15 amperes. How many watts of power are used? _____

2. A 500-watt electric iron is connected to 115 volts. How many amperes, to the nearer hundredth, of current does the iron take? _____

3. Find the current, to the nearer thousandth ampere, used by the lamp shown. _____

4. A one-horsepower motor uses 900 watts when connected to a circuit. The motor receives a current of 3.4 amperes. Find the voltage, to the nearer tenth, across the circuit. _____

5. A heater, rated at 7.5 amperes, is connected to a 220-volt circuit. Find the power, in watts, for the heater. _____

6. How many watts will a motor use if the voltage (E) is 220 volts and the motor is drawing 40 amperes of current? _____

7. What voltage, to the nearer hundredth volt, is necessary to cause a current of 1.74 amperes to exist in a 200-watt lamp? _____

8. A circuit has a voltage of 115 volts and a current of 7.5 amperes. Find the amount of power used. _____

9. What voltage, to the nearer hundredth volt, is necessary for a circuit of 1 625 watts with a current of 14.3 amperes? _____

10. A 550-watt electric iron maintains a voltage of 112 volts. What current, to the nearer hundredth ampere, exists in the iron? _____

11. The resistance coil of the 400-watt soldering iron is shown. Find the current, to the nearer hundredth ampere, received by the resistance coil of the soldering iron. _____

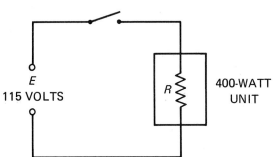

12. A circuit uses 216.2 watts and draws a current of 1.88 amperes. Find the resistance. Express the answer to the nearer hundredth ohm. _____

$$E = \frac{P}{I} \text{ and } R = \frac{E}{I}$$

13. What is the total resistance in a circuit having a voltage of 120 volts and a power of 500 watts? Express the answer to the nearer tenth ohm. _____

$$I = \frac{P}{E} \text{ and } R = \frac{E}{I}$$

14. A motor uses 5 595 watts. Find the horsepower load of the motor. _____

$$hp = \frac{P}{746}$$

15. What is the power in watts in a circuit drawing 59 amperes of current with a total line resistance of 0.67 ohm? _____

$$P = I^2 \times R$$

16. Find the number of watts in the armature of a generator having an internal resistance of 0.3 ohm, and a voltage of 6 volts across the resistance. _____

$$P = \frac{E^2}{R}$$

# Unit 38 COMBINED PROBLEMS ON FORMULAS

BASIC PRINCIPLES OF COMBINED PROBLEMS ON FORMULAS

This unit provides practical problems involving combined problems on formulas.

PRACTICAL PROBLEMS

$$R = \frac{E}{I} \qquad I = \frac{E}{R} \qquad E = IR$$

1. How much current (I) will exist in a lamp that has a resistance of 100 ohms and is connected across a 210-volt circuit? _____

2. What voltage (E) will be required to create a current of 5 milliamperes through a wire-wound resistor having a resistance of 750 ohms? _____

3. An electric heater radiates sufficient heat with a 12-ampere load on a 240-volt circuit. What is the hot resistance? _____

4. What voltage is necessary if 1.5 amperes exist in a 60-watt lamp? _____

$$P = E \times I$$

5. A circuit is connected to a voltage of 120 volts and has a current of 8 milliamperes. How many watts are used? _____

6. A 1 500-watt electric iron is connected to 120 volts. How many amperes will the iron take? _____

7. What is the length of a copper wire 5 mils in diameter, having a resistance of 13.6 ohms? Round the answer to the nearer tenth. _____

$$R = \frac{K \times L}{d^2} \qquad \text{where } \begin{aligned} R &= \text{resistance in ohms} \\ L &= \text{length in feet} \\ d &= \text{diameter in mils} \\ K &= 10.8 \end{aligned}$$

8. What current will exist in a 3 000-watt electric iron when a voltage of 240 volts is maintained? _____

$$I = \frac{P}{E}$$

9. What is the current in a circuit having a resistance of 5 000 ohms if it is consuming 1 500 watts? Round the answer to the nearer thousandth. _____

$$I = \sqrt{\frac{P}{R}}$$

10. A cell has a voltage of 2.1 volts, an internal resistance of 0.9 ohm, and is connected to an electromagnet with a resistance of 0.3 ohm. What current, in amperes, will the electromagnet receive? _____

$$I = \frac{E}{R + r}$$

where E = volts
 I = amperes
 R = resistance of the electromagnet
 r = internal resistance of the cell

11. The impressed voltage across a motor armature is 440 volts and the armature resistance is 100 ohms. If the counter voltage equals 10 volts, how many amperes does the armature take? _____

$$I = \frac{E_x - E_c}{R}$$

where I = amperes
 R = armature resistance
 E_x = impressed voltage in volts
 E_c = counter voltage in volts

12. The temperature of a commutator is 20 °C. Express this temperature in degrees Fahrenheit. _____

$$°C = \frac{(°F - 32°) \, 5}{9}$$

where °F = degrees Fahrenheit
 °C = degrees Celsius

13. What is the amount of wattage used in a single-phase circuit drawing 7.2 amperes, at 110 volts with a 0.8 power factor? _____

$$P = E \times I \times PF \qquad PF = \text{power factor}$$

14. A motor takes a current of 8 amperes per leg on a 440-volt, 3-phase circuit. The power factor is 0.75. What is the load in watts? _____

$$P = \sqrt{3} \times E \times I \times PF$$

15. If a 5-horsepower motor running at full load is connected to a 220-volt circuit, how much current will it take if the motor efficiency is 80%. Round the answer to the nearer thousandth. _____

$$hp = \frac{I \times E \times Eff}{746}$$

where hp = horsepower
 I = current in amperes
 E = voltage in volts
 Eff = percent efficiency

16. What is the impedance of a coil with 84 ohms resistance and 500 ohms reactance? Round the answer to the nearer ohm. _____

$Z = \sqrt{R^2 + X^2}$ where Z = impedance
                                        R = resistance
                                        X = reactance

17. What is the impedance of a circuit having a voltage of 110 volts across it, and a current of 0.025 ampere through it? _____

$Z = \dfrac{E}{I}$ where Z = impedance
                         E = voltage in volts
                         I = current in amperes

Trigonometry

SECTION 9

Unit 39 PYTHAGOREAN THEOREM

BASIC PRINCIPLES OF THE PYTHAGOREAN THEOREM

The Pythagorean thereom states that the sum of the square of the two sides of a right triangle equals the square of the hypotenuse. The *hypotenuse* is the side opposite the right angle.

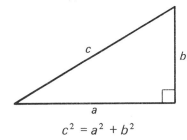

$$c^2 = a^2 + b^2$$

If the value of c is to be found, the formula can be expressed as

$$c = \sqrt{a^2 + b^2}$$

This is done by finding the square root of each side of the equation.

The following formulas can be used to find the other sides of the triangle if the hypotenuse and one side is known.

$$a^2 = c^2 - b^2 \quad \text{or} \quad a = \sqrt{c^2 - b^2}$$
$$b^2 = c^2 - a^2 \quad \text{or} \quad b = \sqrt{c^2 - a^2}$$

PRACTICAL PROBLEMS

Express the answer to the nearer hundredth, when necessary.
Note: Use this illustration for problems 1-6.

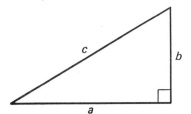

RIGHT TRIANGLE

$$c^2 = a^2 + b^2$$

124

1. $a = 12$ $b = 15$ Find c. _____
2. $a = 24$ $c = 36$ Find b. _____
3. $c = 60$ $b = 30$ Find a. _____
4. $a = 28$ $b = 18$ Find c. _____
5. $c = 48.25$ $b = 22.75$ Find a. _____
6. $a = 12\ 3/4$ $c = 32\ 1/4$ Find b. _____
7. The antenna pole shown is 60 feet high and the guy wire is 5 feet below the top of the pole. What is the length of the guy wire if it is to be fastened 22 feet from the base of the antenna pole? _____

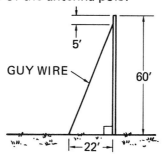

8. The 24-foot ladder is used to make a service entrance connection on a building. The bottom of the ladder cannot be more than 6 feet from the building. What is the minimum height at which the ladder will touch the building? _____

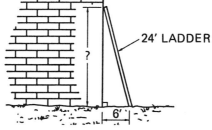

9. The 15-mile power line shown can be straightened out with a new right of way. How long will the new line be? _____

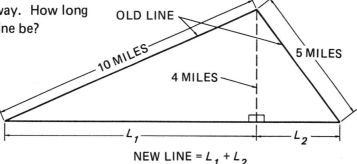

Unit 39 Pythagorean Theorem 125

Unit 40 TRIGONOMETRIC FUNCTIONS

BASIC PRINCIPLES OF TRIGONOMETRIC FUNCTIONS

The angles of a right triangle are determined by the relationship of its sides. The longest side of a right triangle is known as the hypotenuse. The other two sides are known as the opposite side and the adjacent side. The side which is opposite or adjacent is determined by which angle is to be found. The opposite side is the side of the triangle directly opposite the angle to be found. The adjacent side is a side of the angle to be found. The hypotenuse is always opposite the right angle.

Since trigonometric functions are determined by the relationship of the length of the sides, the following formulas can be used to determine the values for sine, cosine, and tangent of the angles.

$$\text{sine of an angle} = \frac{\text{opposite side}}{\text{hypotenuse}}$$

$$\text{cosine of an angle} = \frac{\text{adjacent side}}{\text{hypotenuse}}$$

$$\text{tangent of an angle} = \frac{\text{opposite side}}{\text{adjacent side}}$$

PRACTICAL PROBLEMS

Round the answer to the nearer tenth.
Note: Use this illustration for problems 1-6.

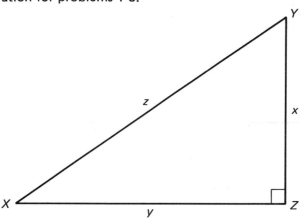

1. z = 10 metres y = 7 metres   Find ∡ X.   _____
2. z = 47 metres x = 23 metres   Find ∡ X.   _____
3. y = 5.42 metres x = 3.3 metres   Find ∡ X.   _____

126

4. y = 13.5 centimetres x = 21.6 centimetres   Find ∡ X. _____
5. x = 15 metres   ∡ Y = 27°   Find z. _____
6. y = 33 metres z = 34.7 metres   Find ∡ Y. _____
7. A roof on a storage shed is 7.3 metres wide with an incline of 20° for drainage. What is the height of the roof? _____

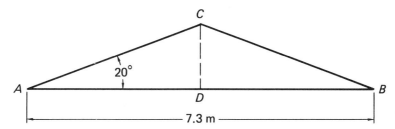

8. A loading ramp 8 metres long is placed from ground level to the top of a loading platform. The platform is 1.5 metres high. What is the angle of incline? _____

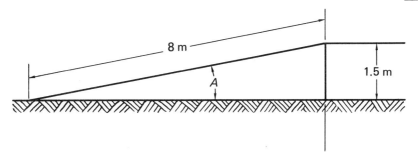

9. An electrician is to bend a pipe to make a 2.2 metre rise in a 5.85 metre distance.

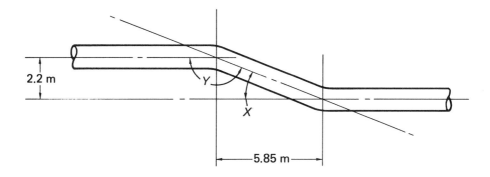

a. Find ∡ X.   a. _____
b. Find ∡ Y.   b. _____

10. A furnace 2 metres wide and 1.5 metres high is placed in an attic with the front side vertically in line with the roof apex. The gabled roof has an incline of 22 degrees and covers a width span of 12 metres. What is the clearance between the top back edge of the furnace and the roof? _____

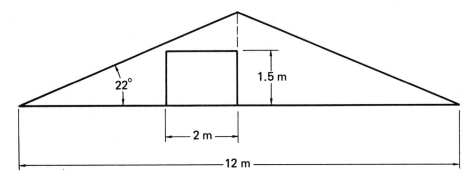

Unit 41 PLANE VECTORS

BASIC PRINCIPLES OF PLANE VECTORS

A *vector* is a line graphically representing any quantity that has both magnitude and direction. Voltages and currents of electricity are directed quantities and can be expressed as vectors. Two or more vectors may be used to represent quantities present at the same instant of time. The *resultant* of forces acting together is a single force that is equivalent to these forces.

- The resultant of quantities having the same direction or phase relationship is found by adding the quantities.

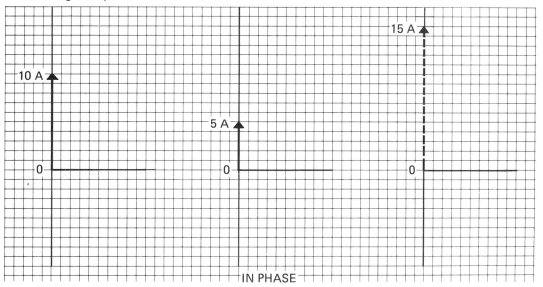

- The resultant of quantities having opposite direction, or 180° out of phase, is found by subtracting the quantities.

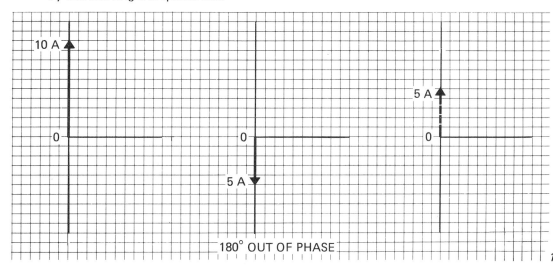

- The resultant of quantities which are not parallel is found to be the quantity represented by the diagonal of the parallelogram of which the two adjacent sides represent the quantities both in magnitude and direction.

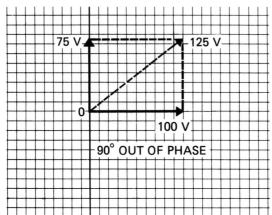

$$OR = \sqrt{V_1^2 + V_2^2}$$
$$OR = \sqrt{(75\text{ V})^2 + (100\text{ V})^2}$$
$$OR = 125\text{ V}$$

where OR = resultant
V_1 = first vector
V_2 = second vector

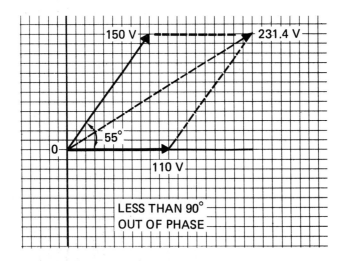

$$OR = \sqrt{V_1^2 + V_2^2 + 2(V_1)(V_2)\cos\theta}$$
$$OR = \sqrt{(110\text{ V})^2 + (150\text{ V})^2 + 2(110\text{ V})(150\text{ V})(0.573\ 6)}$$
$$OR = 231.4\text{ V}$$

where OR = resultant
V_1 = first vector
V_2 = second vector
θ = phase angle

Unit 41 Plane Vectors 131

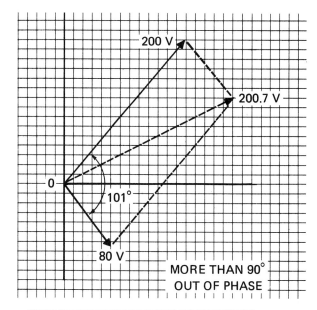

$$OR = \sqrt{V_1^2 + V_2^2 - 2(V_1)(V_2)\cos(180° - \theta)}$$ where OR = resultant

$$OR = \sqrt{(80\text{ V})^2 + (200\text{ V})^2 - 2(80\text{ V})(200\text{ V})(0.190\ 8)}$$ V_1 = first vector

$$OR = 200.7\text{ V}$$ V_2 = second vector

θ = phase angle

Study the electrical impedance diagram.

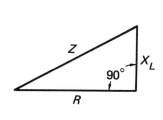

$$Z^2 = R^2 + X_L^2$$

$$Z = \sqrt{R^2 + X_L^2}$$ where Z = impedance

$$X_L = \sqrt{Z^2 - R^2}$$ X_L = reactance

$$R = \sqrt{Z^2 - X_L^2}$$ R = resistance

Study the voltage diagram.

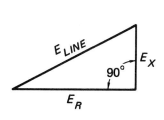

$$E_{LINE} = \sqrt{E_R^2 + E_X^2}$$ where E_{LINE} = line voltage

$$E_R = \sqrt{E_{LINE}^2 - E_X^2}$$ E_R = voltage across the resistance

$$E_X = \sqrt{E_{LINE}^2 - E_R^2}$$ E_X = voltage of the reactance

PRACTICAL PROBLEMS

Express the answer to the nearer hundredth, when necessary.

Note: Use this illustration for problems 1-3.

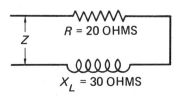

1. Find the impedance Z of the circuit shown if the reactance X_L is 30 ohms, and the resistance R is 20 ohms.

2. If the resistance is changed to 40 ohms, what is the impedance?

3. The impedance Z is 20 ohms and the resistance R is 10 ohms. What is the reactance X_L?

4. What is the value of the resistance (R) in the series circuit shown?

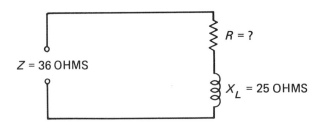

Note: Use this diagram for problems 5-7.

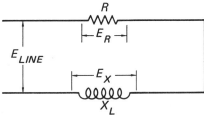

5. Voltage E_R in the voltage diagram is 14 volts, and voltage E_X is 18 volts. What is the line voltage E_{LINE}?

6. The line voltage E_{LINE} is 120 volts and the voltage of the reactance E_X is 60 volts. What is the voltage across the resistance E_R?

7. The voltage E_X is 150 volts and E_R is 200 volts. Find the voltage across the line E_{LINE}?

8. In the circuit shown E_c is 135 volts and E_b is 175 volts. Find E_a. _____

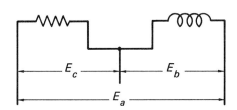 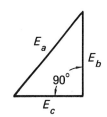

VOLTAGE DIAGRAM

Note: Use this illustration for problems 9-10.

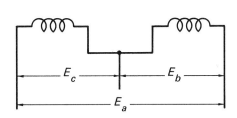 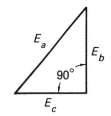

VOLTAGE DIAGRAM

9. This diagram represents a 3-wire, 2-phase voltage source. If E_b is 130 volts and E_a is 182 volts, find E_c. _____

10. Determine the voltage of E_a if E_b is 105 volts and E_c is 120 volts. _____

11. A motor draws 3.8 amperes of current when operated from a 120 volt, 60 hertz power source. The motor winding has a series resistance of 30 ohms and an inductive reactance of 10 ohms. Find the line voltage to the nearer whole volt. _____

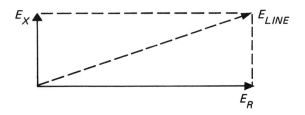

12. A transformer secondary circuit is shown. Find I_s. Express the answer to the nearer tenth. _____

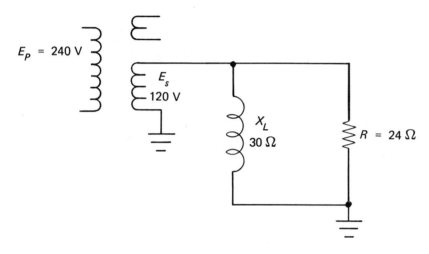
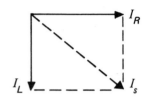

13. A 20 ohm resistor and a capacitor of 15 ohms capacitive reactance at 60 hertz are connected in parallel across a 60 volt, 60 hertz source. Find I_{LINE} to the nearer tenth. _____

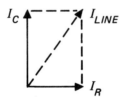

14. Two voltages, E_1 and E_2 are applied to a resistor. E_1 is 40 volts at 0° reference. E_2 is 60 V and leads E_1 by 40°. Find E_R to the nearer tenth. _____

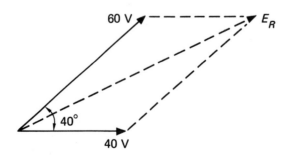

Unit 42 ROTATING VECTORS

PRINCIPLES OF ROTATING VECTORS

The concept of vectors rotating in a circular motion is used in the study of alternating current and voltage. The vector is the radius of a circle and is rotated in a counter-clockwise direction at a constant speed. When the vector is rotated from **A** to **B**, the angle between the reference axis and the vector position at **B** is **45°**. At position **C** the angle is **90°**.

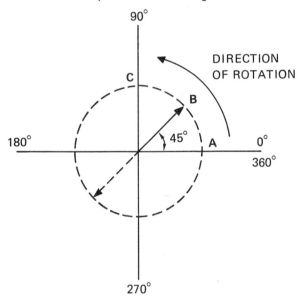

- The rotating vector may be resolved into its vertical component for each degree of rotation or any instant in time. The magnitude of the vertical component is plotted as a sine curve.

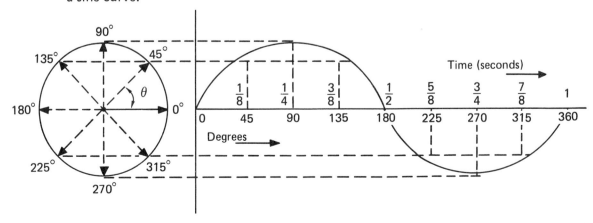

135

136 Section 9 Trigonometry

- An angular rotation of 360° is one *cycle*. The *frequency* is the number of complete cycles per second. The horizontal reference axis of the rotating vector is extended to the right to form the horizontal time axis for the curve.

Study the formula for finding frequency.

$$f = \frac{1}{t}$$

where f = frequency in hertz

t = time in seconds for one rotation

Study the formula for finding instantaneous voltage.

$$E_i = E_{max} \times \sin\theta$$

where E_i = instantaneous voltage

E_{max} = maximum voltage

θ = angle of rotation

Study the formula for finding instantaneous current.

$$I_i = I_{max} \times \sin\theta$$

where I_i = instantaneous current

I_{max} = maximum current

θ = angle of rotation

The voltage curves of two AC generators 90 degrees out of phase are illustrated. The rotating vectors to the left of the voltage curves represent the AC generators. Generator 2 is started one-fourth of a cycle after generator 1. Their generated voltages are separated by the period of time indicated as 90°.

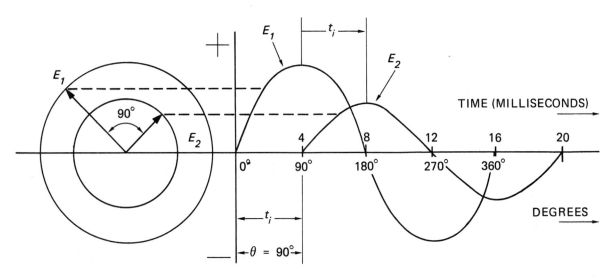

Study the formula for finding the phase angle.

$$\theta = t_i \times f \times 360$$

where θ = phase angle

t_i = time interval, in seconds between the two voltages

f = frequency of the AC generators

360 = number of degrees in one cycle

PRACTICAL PROBLEMS

1. An AC current has a peak value of 50 amperes. Find the instantaneous value of the current at 10 degrees. Express the answer to the nearer hundredth. _____

2. What is the magnitude of voltage at point A? Express the answer to the nearer tenth. _____

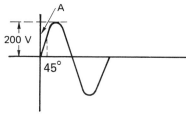

3. An AC voltage wave has an instantaneous value of 80 volts at 30 degrees. Find the maximum voltage value of the wave. _____

4. An AC voltage has an instantaneous value of 110 volts. Find, to the nearer degree, the phase angle if the peak voltage is 155.5 volts. _____

5. An AC current has a value of 18 amperes at 12 degrees. What is the value at 90 degrees? Round the answer to the nearer hundredth. _____

6. Find the phase angle at which an instantaneous voltage of 144 volts appears in a wave with peak value of 500 volts. Round the answer to the nearer degree. _____

7. The voltage of an AC wave is 100 volts at 30 degrees. Find the instantaneous voltage at each value.
   a. 0°   a. _____
   b. 15°   b. _____
   c. 60°   c. _____
   d. 75°   d. _____
   e. 90°   e. _____

8. What is the time in milliseconds required to generate one cycle of voltage at 60 hertz?

9. In 10 microseconds a generator produces one hertz of voltage. What is the frequency of the generator?

10. What is the phase angle between two 60 hertz voltage generations separated by a time interval of three milliseconds?

Unit 43 COMBINED PROBLEMS ON TRIGONOMETRY

BASIC PRINCIPLES OF COMBINED PROBLEMS IN TRIGONOMETRY

This unit provides practical problems in trigonometry. Use the formulas provided in previous units to solve these problems.

PRACTICAL PROBLEMS

Solve problems 1-5 for the remaining part of the right triangle shown in the figure. Round the answer to the nearer hundredth.

1. If a = 4 b = 8
2. If b = 25 c = 40
3. If c = 350 a = 200
4. If a = 7.5 b = 12.5
5. If b = 12.4 c = 67.8

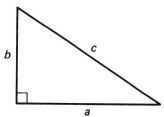

Using the impedance diagram, solve for the missing quantities in each series circuit for problems 6-10. Round the answer to the nearer tenth.

Z = impedance
R = resistance
X_L = inductive reactance

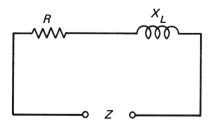
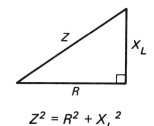

$$Z^2 = R^2 + X_L^2$$

6. If X_L = 500 ohms R = 500 ohms
7. If R = 45 ohms Z = 90 ohms
8. If Z = 1 100 ohms X_L = 400 ohms
9. If the inductive reactance is 7.2 ohms, and the resistance is 10.2 ohms, what is the impedance? Express the answer to the nearer tenth.
10. In the series circuit, if the impedance is 70 ohms and the reactance is 50 ohms, what is the value of the resistance? Round the answer to the nearer ohm.

Use this diagram for problems 11-12. Round the answers to the nearer tenth.

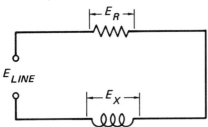
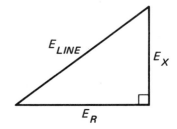

11. What is the voltage across the reactance E_X if E_{LINE} = 240 volts, and E_R = 120 volts? _____

12. If the voltage across the reactance E_X is 50 volts and the line voltage E_{LINE} is 300 volts, what is the voltage across the resistance E_R? _____

13. What is the resultant of two vectors if OA = 7 amperes, OB = 10 amperes, and the phase angle is 50°? Round the answer to the nearer tenth. _____

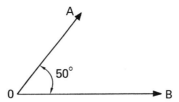

14. What is the output frequency, to the nearer ten hertz, of a generator that produces two voltage waveforms in 5.55 milliseconds? _____

15. What is the phase difference, in degrees, between the voltages across R_L if generator #1 is started 2.77 milliseconds before generator #2? The frequency of both generators is 120 hertz. Round the answer to the nearer degree. _____

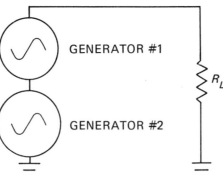

16. What is the maximum peak value of a 60 hertz wave that has an instantaneous voltage of 22 volts 1.67 milliseconds before the end of one cycle? Round the answer to the nearer tenth. _____

ACHIEVEMENT REVIEW A

The following testing material is provided for the convenience of the instructor. Delmar Publishers gives permission to reproduce this material in whole or in part to meet the individual needs of the instructor.

Note: The numbers in parentheses, ( ), given below each problem show the unit in which similar problems have been discussed. Formulas needed for these problems are found in the appendix.

1. An electrician purchases 553 octal boxes for a number of jobs. On seven of the jobs 63, 56, 91, 79, 74, 57, and 68 boxes are used. How many are left for the last job?
   (1, 2) _____

2. A contractor purchases a 606-foot reel of cable. For one electrical project, the following quantities are used: three 12-foot pieces, seven 22-foot pieces, and five 17-foot pieces. How many feet remain on the reel?
   (1, 2, 3) _____

3. A total load of 49 203 watts is distributed equally over 11 branch circuits. Find the average load per circuit.
   (4) _____

4. A motor uses 2.8×10^3 watts from a 4.4×10^2 volt line. Find, to the nearer thousandth ampere, the current drawn.
   (28, 37) _____

5. The following fractional horsepower motors are used in a plant: 1/16 hp, 1/8 hp, 1/3 hp, 1/4 hp, 3/4 hp, and 1/2 hp. What is the combined horsepower of all the motors?
   (6) _____

6. An electrician uses the following amounts of EMT: 9.004 metres, 0.07 metre, 7.137 metres, 6.879 4 metres, and 578.9 metres. Find the total number of metres of EMT used.
   (11) _____

7. The total current in amperes in a lighting circuit is equal to the sum of the lamp currents. The following lamps are in a circuit: two 25-watt (0.208 ampere each); three 100-watt (0.834 ampere each); two 40-watt (0.334 ampere each). Find the total current when all lamps are on.
   (11, 13) _____

8. A pipe has a wall thickness of 0.37 centimetre and an outside diameter of 8.22 centimetres. What is the inside diameter?
(12, 13, 25)

9. An armature shaft that is 1 3/8 inches in diameter is found to have an irregular surface. It is decided that 0.017 inch will be taken off the surface. Express the new diameter in decimal form.
(12, 13, 15)

10. Standard 4-inch conduit has an inside diameter of 4.026 inches and an outside diameter of 4 1/2 inches. Find the thickness of the wall.
(12, 14, 15)

11. What is the difference between the diameters of No. 1 AWG wire of 7.348 2 centimetres and No. 7 wire of 3.675 4 centimetres?
(12)

12. What voltage is necessary to operate a 36-watt trouble light, if the lamp has a hot resistance of 4 ohms?
(37)

13. A panel 23 inches long has 5 holes spaced equally across in a straight line. Each hole is 1.73 inches in diameter, and the distance from the edge of the panel to the first and last hole is equal to the distance between the edges of the consecutive holes. What is the spacing between the holes to the nearer thousandth inch?
(12, 13, 14)

14. Light fixtures for six rooms cost $14.35, $22.65, $18.43, $19.95, $23.15, and $17.49. What is the average cost per room for fixtures?
(11, 14, 21)

15. An electrician purchases various materials for a job. The list price for the materials is $945.52, but the electrician receives discounts of 15% and 8%. How much does the electrician pay?
(19)

16. An entrance switch lists at $89.30. Find the net price with discounts of 30% and 5%.
(19)

17. A 120-volt circuit has a resistance of 54.3 ohms and a current of 2.2 amperes. How many watts of power are used?
(37)

18. In a 240-volt circuit a resistance of 30 ohms takes 8 amperes of current. Find the amount of power used.
(37)

19. Find the total number of volts, to the nearer thousandth, in a circuit having a power of 743 watts and a resistance of 9 ohms.
(23, 37)

20. A wire carries a current of 0.078 ampere and has a cross-sectional area of 1 024 circular mils. Find the diameter of the wire.
(23, 35)

21. A copper wire 237 feet long has a resistance of 1.107 ohms. What is the resistance, to the nearer thousandth, of 862 feet of the same kind of wire?
(31, 35)

22. Solve the following expression for L:

$$R = \frac{KL}{d^2}$$

(34)

23. In an electrical circuit, R_1 = 20 ohms and R_2 = 30 ohms. The total voltage E is 120 volts. Find the total current I, in amperes.

$$R_T = \frac{R_1 \times R_2}{R_1 + R_2} \qquad I = \frac{E}{R_T}$$

(35, 36)

24. In a circuit R_1 equals 20 ohms, R_2 equals 40 ohms, and R_3 equals 50 ohms. Use the formula shown to find the total resistance, R_T, of the circuit. Round the answer to the nearer thousandth.

$$R_T = \frac{1}{\frac{1}{R_1} + \frac{1}{R_2} + \frac{1}{R_3}}$$

(35)

25. Express 2.2 microfarads as farads.
(28)

26. A room in an industrial plant has 70 fluorescent lamps each with a rating of 60 watts. The size of the room is 52 feet by 47 feet. Find the number of watts per square foot of area. Express the answer to the nearer thousandth.
(26)

27. Impedance Z in a series circuit equals $\sqrt{R^2 + X^2}$. If R equals 80 ohms and X equals 60 ohms, find Z.
(41)

28. If $Z^2 = R^2 + X_L^2$, solve the equation for X_L.
    (34)

29. In the right triangle shown, solve for E_T.

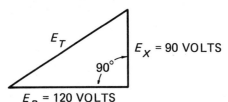

30. A carbon brush is 1.73 inches in width, 3/4 inch in height, and 1 13/16 inches in length. If 1/8 inch is taken off the width, what is the new dimension for the width?
    (12, 15)

31. Find the inductive reactance of a coil if the frequency is 10 kilohertz and the inductance is 10 millihenrys. Use the formula $X_L = 2\pi f L$ where $\pi = 3.141\,6$.
    (28, 35)

32. An 80 metre long wire has a resistance of 9.25 ohms. What is the resistance, to the nearer thousandth, of 1 metre?
    (31)

33. How many watts of power are lost in a transformer rated at 850 kilowatts if 6 percent of the energy is lost as heat?
    (17, 28)

ACHIEVEMENT REVIEW B

The following testing material is provided for the convenience of the instructor. Delmar Publishers gives permission to reproduce this material in whole or in part to meet the individual needs of the instructor.

Note: The numbers in parentheses, ( ), given below each problem show the unit in which similar problems have been discussed. Formulas needed for these problems are found in the appendix.

1. An electrician collected $437 for an electrical renovation job. For the various materials used on the job, the electrician pays $84, $37, $13, and $77. How much is received for labor?
   (1, 2) _____

2. For a certain room, the maximum number of watts permitted is 1 260. The room contains six 40-watt lamps, four 75-watt lamps, and three 100-watt lamps. How many 60-watt lamps can be placed in the room?
   (1, 2, 3, 4) _____

3. Four electricians work a total of 840 hours on a job. Each works a schedule of 7 hours per day, 5 days per week. How many weeks does each work?
   (3, 4) _____

4. A concrete pad 1.75 metres by 3.2 metres supports three transformers. Each transformer base measures 75 centimetres by 90 centimetres. Find the percent of the pad surface area covered by the transformers. Round the answer to the nearer hundredth.
   (17, 26) _____

5. How many 13 1/16-inch pieces can be obtained from a fiber strip that is 7 feet, 7 7/16 inches long?
   (9, 25) _____

6. An electrician travels 534 kilometres in 8 hours to get to a job. At the same rate of speed find the number of kilometres traveled in 12 hours.
   (31) _____

7. If a motor rotates at the rate of 347 1/4 revolutions per minute, how many revolutions will it make in 1/3 hour?
   (8) _____

8. An electrical firm borrows $9 347 to purchase motors for five jobs. The loan was made at an interest rate of 12% per annum, and it was paid back in 14 months. What was the total amount paid back?
(18) _____

9. A 120-volt circuit contains a resistance of 40 ohms and uses 360 watts of power. Find the current.

$$I = \sqrt{\frac{P}{R}}$$

(37) _____

10. The turns ratio of a transformer is 12:5. Find the inverse ratio.
(30) _____

11. The circumference of a circle equals the diameter times π. Find the difference in circumference between a 6.34-inch pulley and a 9 1/16-inch pulley. Use π = 3.141 6 and round the answer to the nearer thousandth.
(35) _____

12. Express 3/10 000 as a number between 1 and 10 times the proper power of ten.
(28) _____

13. Standard 6-inch conduit has an inside diameter of 6.065 inches and an outside diameter of 6 5/8 inches. Find the wall thickness.
(15, 25) _____

14. A motor rated at 45 horsepower is actually developing 52 horsepower. What is the percent of horsepower overload? Express the answer to the nearer hundredth percent.
(17) _____

15. If $327.60 is the amount of profit on a job, and this represents 12% of the contract price, what is the contract price?
(17) _____

16. An electrician installs a bell system consisting of seven 24-volt, 50-milliampere bells connected in parallel. What is the total power used by this system?
(37) _____

17. If the capacitive reactance X_c is 53 ohms at a frequency f of 60 hertz, find the value to the nearer microfarad of the capacitor C.

$$X_c = \frac{1}{2\pi fC}$$

(35) _____

18. One square foot equals approximately 0.093 square metre. Find the floor area of a building 44 1/4 feet long and 14 3/4 feet wide to the nearer hundredth square metre.
    (26)

19. Power, in watts, is equal to E^2/R. A 120-volt circuit uses 900 watts of power and draws 7.5 amperes of current. Find the resistance to the nearer ohm.
    (37)

20. Voltage, in volts, is equal to $\sqrt{\text{Power} \times \text{Resistance}}$. A circuit uses 605 watts of power, and has a current of 5.5 amperes for a 20-ohm resistance. What is the voltage?
    (23, 37)

21. Find the number of inches in 45.72 centimetres. Use 1 inch equals 2.54 centimetres.
    (25)

22. A junction box, 7 inches wide and 12 7/16 inches high, is installed. Find the number of square inches of wall area needed for the box.
    (26)

23. Solve the equation for r.
    $$I = \frac{E \times N}{r \times N + R}$$
    (34)

24. The area of a circle is approximately equal to 0.785 4 times the diameter squared. Find the area, to the nearer thousandth square inch, of a metal plate that is 5 3/4 inches in diameter if it has a 1 1/16-inch diameter hole in it.
    (26)

25. If a 15-horsepower motor is connected to a 220-volt line, how much current, to the nearer thousandth, will the motor take if the motor efficiency is 92%?

    $$hp = \frac{I \times E \times Eff}{746}$$ where E = volts
    hp = horsepower
    I = amperes
    Eff = motor efficiency
    (35)

26. Two wire NM cable is used to connect four outlets. The following wire lengths are needed to complete the job: 6.2 metres, 7.9 metres, 19.5 metres, and 7.3 metres. What is the total length of wire used?
    (11)

148 Achievement Review B

27. A school study hall is 12 metres wide by 20 metres long. Two rows of lighting are placed in the ceiling, lengthwise. What is the center-to-center distance between the rows of lights if the rows are spaced with equal distances from the side walls and between the rows?
(4)

28. The total amperes in an electrical circuit is equal to the sum of the amperes taken by each lamp in the circuit. In a certain circuit, 12 lamps require a total current of 11.76 amperes. Each lamp in the circuit takes the same amount of current. What is the amount of current required for 5 lamps?
(13, 14)

29. An antenna pole is 47 feet above the ground, and a guy wire is attached 7 feet below the top of the pole. What is the length of the guy wire if it is to be fastened to the ground 30 feet from the base of the antenna pole?

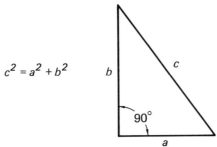

(39)

30. Since $Z^2 = R^2 + X^2$, find the value of R if $Z = 1\,200$ ohms, and $X = 600$ ohms. Round the answer to the nearer ohm.
(41)

31. The impedance Z, of a series RC circuit, equals $\sqrt{R^2 + X_c^2}$. If $Z = 50$ ohms and $X_c = 30$ ohms, find R.
(41)

32. A ladder 7.3 metres long is placed against a wall 1.2 metres from the base. What angle does the ladder make with the ground? Round the answer to the nearer degree.
(40)

33. A loading platform is 1.4 metres high. A loading ramp is placed from the platform top to ground level with an incline of 15 degrees. What is the length of the ramp to the nearer tenth?
(40)

 # DIAGNOSTIC READING SURVEY

To The Teacher: Detailed instructions for administering this survey and interpreting the results are provided in the Instructor's Guide. Please follow them exactly.

To The Student: This survey is to see what basic skills you already have and what skills you may need to improve in this program area. The information gained from this survey will help your instructor plan the kind of instruction that is best for you.

This is a timed survey. If you complete a section before time is called, you may not go on to the next section and you may not return to a previous section. You may go over your answers in the section you have just completed. When you are sure of your answers, turn your answer sheet over. If everyone completes a section before the time limit, the timing may be stopped.

The first section is Vocabulary. This section measures how well you know the meaning of words. It contains ten items and you will have ten minutes to complete your work.

The second section is Spelling. This section contains ten items and you will have five minutes to complete your work.

The third section is Comprehension and will measure your knowledge of reference information and how well you understand what you read. It contains twenty-five items and you will have forty-five minutes to complete your work.

The complete survey takes one hour. Follow all directions carefully. Write answers neatly in the correct spaces on your answer sheet.

This reading survey has been especially prepared by Carla Hoke for students interested in electricity and electronics.

The Competency-based Reading Survey is designed to determine skill levels in the following areas: Vocabulary, Spelling, and Comprehension. The subskills in the Comprehension section include Reference Skills, Using a Table of Contents and an Index, Finding the Main Idea, Locating Facts, Following Directions, and Reading Charts and Diagrams.

Item analysis sheets will provide specific information on each student's reading strengths and weaknesses and will aid both the reading specialist and the vocational instructor in planning developmental and corrective programs and in individualizing instruction.

Each student will need:

- the Electricity and Electronics Reading Survey
- an answer sheet
- a pencil or pen

VOCABULARY

DIRECTIONS: Look at the sample test word below. The word is *battery*. Now read the definitions just below *battery*. Find the one definition in this group that means most nearly the same as *battery*. Definition "b" is the correct definition, "an electro-chemical device for producing electricity." On your answer sheet, under the section marked Vocabulary, find the line marked "Sample" and write a "b" on that line.

For each numbered word in this section, write the letter of the definition that means most nearly the same as the given word on your answer sheet. If you are not sure of an answer, make the most careful guess you can.

Sample: battery

      a. a replaceable cylinder

      b. an electrochemical device for producing electricity

      c. device found only in a car or truck

      d. device for changing electrical energy into chemical energy

      e. none of the above

STOP

Do not turn the page until you are told to begin.

1. volt
    a. practical unit of resistance
    b. practical unit of electric current
    c. practical unit of electromotive force
    d. practical unit of electrical power
    e. none of the above

2. current
    a. flow of electrons
    b. popular
    c. instrument used to measure heat
    d. unit of measurement
    e. none of the above

3. ammeter
    a. unit of length
    b. instrument used to measure resistance
    c. unit of weight
    d. instrument used to measure current
    e. none of the above

4. generator
    a. device found only in a car or truck
    b. device that converts mechanical energy into electrical energy
    c. an instrument which measures heat
    d. device that changes electrical energy into mechanical energy
    e. none of the above

5. resonance
    a. property of batteries
    b. ability of material to produce sound
    c. ability of material to produce electricity
    d. that adjustment of a current which allows the greatest flow of current of a certain frequency
    e. none of the above

6. Ohm
    a. device used to measure current
    b. unit of measure of resistance
    c. unit of measure of current
    d. instrument used to measure length
    e. none of the above

7. resistance
    a. ability of material to withstand the flow of electrons
    b. ability of material to encourage the flow of electrons
    c. ability of material to convert neutrons to electrons
    d. ability of material to convert protons to electrons
    e. none of the above

8. ground
    a. bottom of a generator
    b. connection of an electrical conductor with the earth
    c. lowest part of a motor
    d. another name for current
    e. none of the above

9. circuit
    a. device which directs the flow of atoms
    b. complete path along which electrons can move
    c. a coil of wire
    d. a type of conductor
    e. none of the above

10. capacitor
    a. instrument to measure amount something can hold
    b. two insulators separated by a conductor
    c. two conductors separated by an insulator
    d. device which measures electric current
    e. none of the above

_ **STOP**

Turn your answer sheet over and wait for further directions.

SPELLING

DIRECTIONS: Look at the sample below. Only one word is correctly spelled. The correct spelling is "c", wiring. On your answer sheet, under the section marked Spelling, find the line marked "Sample" and write a "c" on that line.

For each numbered problem in this section, write the letter of the word which is correctly spelled on your answer sheet. If you are not sure of an answer, make the most careful guess you can.
_ _

Sample:
    a. wireing
    b. wirring
    c. wiring
    d. whiring

_ **STOP**

Do not turn the page until you are told to begin.

1. a. searies
    b. cearies
    c. series
    d. serease

2. a. transmitter
    b. transmiter
    c. transmittor
    d. transmitor

3. a. fuze
    b. fuse
    c. fuese
    d. fueze

4. a. insullation
    b. insulashun
    c. insulasion
    d. insulation

5. a. parralel
    b. pairalel
    c. parallel
    d. pearalil

6. a. splise
    b. splice
    c. splies
    d. splize

7. a. whatt
    b. whadt
    c. watt
    d. wadt

8. a. receptacle
    b. reseptacle
    c. receptacel
    d. reseptakl

9. a. modar
    b. motor
    c. moder
    d. moter

10. a. junktion
    b. juncshun
    c. junksion
    d. junction

STOP
_____

PRACTICE SPACE

COMPREHENSION

DIRECTIONS: This section of the test consists of two parts. The items in the first part measure your knowledge of reference information. The items in the second part show how well you understand what you read.

This section contains 25 items. Choose the best answer for the items in the first part. In the second part, read the information given and do the items that follow. Fill in the space on your answer sheet with the letter of the answer you choose.

Sample: Read the passage below.

Joe decided to get a part-time job after school. He looked in the classified section of the newspaper since this part contains the help wanted notices. He found two companies near home which were hiring and called for an appointment for an interview.

Now read the item below.

The classified section of the newspaper contains

a. secret information

b. help wanted notices

c. current news stories

d. advice to the lovelorn

Look at the section of your answer sheet under the word Comprehension. Put a "b" on the line for the answer for the sample item because the classified section of the newspaper contains "help wanted notices."

STOP

Do not turn the page until you are told to begin.

Items 1-3 measure your knowledge of reference information.
Read each question and mark your answer on the answer sheet.

1. A glossary contains
   a. definitions
   b. quotes from famous people
   c. tables and diagrams
   d. an overview of the book

2. The appendix of a book consists of
   a. unusual words used in the text
   b. a list of chapter headings
   c. a list of authors
   d. supplementary material

3. A bibliography contains
   a. a list of important names and dates
   b. a list of books
   c. a list of topics covered in the text
   d. a list of words used in the text

Now read the Table of Contents. Answer the questions which follow.

TABLE OF CONTENTS	
CHAPTER	PAGE
1. Electricity—The Behavior of Electrons	1
2. Electrical Circuits	22
3. Electrical Energy & Power	41
4. Electric Heating & Lighting	86
5. Cells & Batteries	129
6. Motors	154
7. Amplifiers	221
8. Safety in Electricity-Electronics Shops	235
9. Careers in Electricity & Electronics	253

4. The chapter "Electric Heating & Lighting" begins on page
   a. 41   b. 68   c. 86   d. 129

5. Page 145 is in which chapter?
   a. "Motors"
   b. "Cells & Batteries"
   c. "Electrical Energy & Power"
   d. "Amplifiers"

6. Information on an occupation in electrical engineering can be found in which chapter?
   a. 1   b. 5   c. 7   d. 9

Read the Index on the following page. Then do items 7-9 and mark your answers on the answer sheet.

> **INDEX**
>
> Charges, electric, defined, 3
>   laws of, 4
> Circuit breakers, defined, 126-127
> Clamps, cable, 147
>   in boxes, 161
>   ground, 203
> Codes, local, 92
>
> Conductors, defined, 5
>   insulation, 97-98
>   joints, 111
>   sizes, 93-95
>   skin effects on, 95-96
>   soldering, 110-111

7. The meaning of the word "conductor" is on page
    a. 3
    b. 4
    c. 5
    d. 111

8. The local electrical codes can be found on page
    a. 203
    b. 92
    c. 93
    d. 147

9. The effects of conductors on the skin begins on page
    a. 161
    b. 5
    c. 93
    d. 95

Read the passages which begin on the next page. After each passage, choose the best answer for each item and mark your answer on the answer sheet.

For a rough classification, magnets may be divided into two groups: permanent or temporary. A permanent magnet is intended to keep its atomic arrangement steady after the magnetizing force is removed. Permanent magnets are used in telephone receivers, door latches, small d.c. motors, electrical measuring instruments, magnetos, speedometers, and a great variety of gadgets. For years, high-carbon tool steel, and a few alloy steels (cobalt, molybdenum, chrome-tungsten) were the only useful permanent magnet materials. Later, various alloys were developed, such as Alnico (aluminum, nickel, cobalt, iron), which is widely used as a high-strength permanent magnet. Flexible rubbery magnets are made by mixing magnetic oxide powders with vinyl and other plastics.

Temporary-magnet materials, which are easily and strongly magnetized but lose most of their strength when the magnetizing force is removed, are of more importance than permanent-magnet materials, both in total amount in use and in variety of applications. The first-used material for temporary magnets was plain iron, as pure as could be obtained, softened by annealing (slow cooling). A soft metal is one in which atoms slide around readily, permitting atoms to be magnetically disarranged easily. The most-used material is silicon iron (2-4% Si), a soft alloy used in transformers, most motors and generators, relays and other magnetic equipment built in large quantities.

Answer questions 10-13.

10. The best title for this selection is
    a. "Temporary Magnets"
    b. "Permanent Magnets"
    c. "Types of Magnetic Materials"
    d. "Electromagnetic Force"

11. As used in this passage, the term "annealing" means
    a. fast cooling
    b. slow cooling
    c. easily magnetized
    d. softening

12. Which of the following is not a permanent magnet material?
    a. silicon
    b. high-carbon tool steel
    c. alloy steel
    d. Alnico

13. Temporary magnets are used in
    a. telephone receivers
    b. door latches
    c. speedometers
    d. transformers

Read the passage on the following page and answer questions 14-16.

To use an ohmmeter, the tips of the test leads are first held together (short-circuited) and the rheostat adjusted so that the pointer of the meter moves over to the righthand end of the scale, where the "zero ohms" mark is placed. The meter now indicates that there is zero resistance between the test leads.

When the test leads are separated, there is no current in the circuit, and the pointer drops back to the left end of the scale, where the "infinity ohms" mark is placed.

When the test leads are touched to the ends of a resistor of unknown value, the resistance is read directly from the "ohms" scale. Normally, an ohmmeter has several ranges, with different combinations of series resistance and battery voltage used in each range.

The ohmmeter diagrammed in Fig. 7-10 is called a series ohmmeter, and is usually installed in a case containing multiple-contact switches, voltmeter resistors, and ammeter shunts, forming the "multimeter" or "volt-ohmmeter" that is widely used in testing electronic equipment.

MEASURING INSTRUMENTS

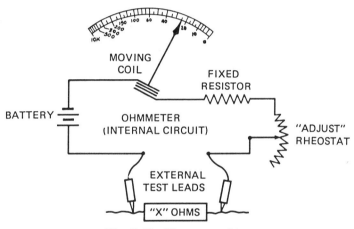

Fig. 7-10 Ohmmeter Circuit

14. To use an ohmmeter, the pointer must first be set at
    a. zero   b. infinity   c. 10 ohms   d. 10 K ohms

15. The following steps are used to measure resistance in a circuit. Which step is taken first?
    a. connect ohmmeter leads together   c. turn off power
    b. discharge electrolytic capacitors   d. read "ohm" scale

16. When there is no current in a circuit, what is the reading on the ohmmeter scale?
    a. zero   b. infinity   c. 10 ohms   d. 100 ohms

Read the passage on the following page and answer questions 17-20. Put your answers on the answer sheet.

Light is energy that is radiated by electronic disturbances in atoms. Electrons in an atom can accumulate energy in many ways: from heat, as in a red-hot object; or by being hit by other electrons, as in a gas-conduction tube; or by absorbing energy radiated by other materials. Sooner or later, this absorbed energy is given out. The amount of energy that an electron can get rid of in one burst depends on where the electron is, that is, what kind of atom it is in, and its location in the atom.

The energy is radiated as a wave-like pulse of electric lines of force and magnetic lines of force, which is called an electromagnetic wave. The vibration frequency of these traveling lines of force is proportional to the amount of energy that the electron gave off. Frequencies in the range from 4.3×10^{14} to 7.5×10^{14} vibrations per second affect electrons in our eyes. We call electromagnetic waves in this frequency range by the name of "light."

Waves of a frequency slightly higher than 7.5×10^{14} are called "ultra-violet," waves in a lower frequency range, from 10^{11} to 10^{14} vibrations per second are called "infra-red," and heat radiation. The term "black light," often used by newspaper reporters who do not know the correct terms, may mean either ultra-violet or infra-red. Ultra-violet and infra-red differ greatly in their effects and uses.

17. The best title for this passage is
    a. "How Light Is Measured"
    b. "What Is Light?"
    c. "Sources of Light"
    d. "Commercial Light Sources"

18. Which of the following is not one of the ways electrons in an atom accumulate energy?
    a. by being hit by other electrons
    b. by absorbing energy radiated by other materials
    c. from cooling temperatures
    d. from heat

19. Define "light"

20. Define "light" a second way

On the following page you will find a diagram. Use the diagram to answer questions 21-23. Mark your answers on the answer sheet.

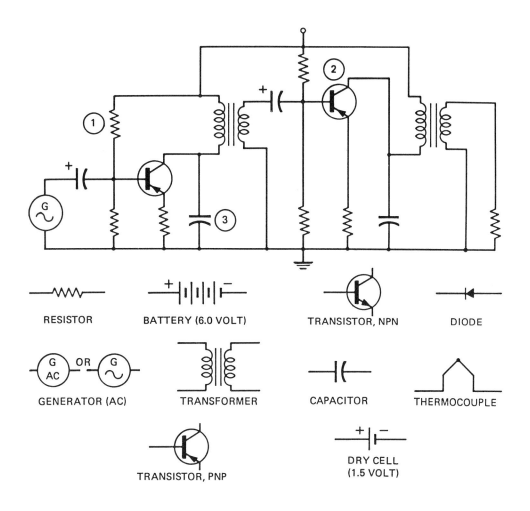

21. Find 1 on the diagram. This symbol refers to a
    a. capacitor
    b. resistor
    c. transformer
    d. battery

22. Find 2 on the diagram. This symbol refers to a
    a. diode
    b. transistor, NPN
    c. transistor, PNP
    d. thermocouple

23. Find 3 on the diagram. This symbol refers to a
    a. battery
    b. generator
    c. dry cell
    d. capacitor

Use the Table of Natural Sines and Cosines to answer questions 24-25.

NATURAL SINES AND COSINES

NOTE: For cosines, use right-hand column of degrees and lower line of tenths.

Deg.	°0.0	°0.1	°0.2	°0.3	°0.4	°0.5	°0.6	°0.7	°0.8	°0.9	°1.0	
45	0.7071	0.7083	0.7096	0.7108	0.7120	0.7133	0.7145	0.7157	0.7169	0.7181	0.7193	44
46	0.7193	0.7206	0.7218	0.7230	0.7242	0.7254	0.7266	0.7278	0.7290	0.7302	0.7314	43
47	0.7314	0.7325	0.7337	0.7349	0.7361	0.7373	0.7385	0.7396	0.7408	0.7420	0.7431	42
48	0.7431	0.7443	0.7455	0.7466	0.7478	0.7490	0.7501	0.7513	0.7524	0.7536	0.7547	41
49	0.7547	0.7559	0.7570	0.7581	0.7593	0.7604	0.7615	0.7627	0.7638	0.7649	0.7660	40
50°	0.7660	0.7672	0.7683	0.7694	0.7705	0.7716	0.7727	0.7738	0.7749	0.7760	0.7771	39
51	0.7771	0.7782	0.7793	0.7804	0.7815	0.7826	0.7837	0.7848	0.7859	0.7869	0.7880	38
52	0.7880	0.7891	0.7902	0.7912	0.7923	0.7934	0.7944	0.7955	0.7965	0.7976	0.7986	37
53	0.7986	0.7997	0.8007	0.8018	0.8028	0.8039	0.8049	0.8059	0.8070	0.8080	0.8090	36
54	0.8090	0.8100	0.8111	0.8121	0.8131	0.8141	0.8151	0.8161	0.8171	0.8181	0.8192	35
55	0.8192	0.8202	0.8211	0.8221	0.8231	0.8241	0.8251	0.8261	0.8271	0.8281	0.8290	34
56	0.8290	0.8300	0.8310	0.8320	0.8329	0.8339	0.8348	0.8358	0.8368	0.8377	0.8387	33
57	0.8387	0.8396	0.8406	0.8415	0.8425	0.8434	0.8443	0.8453	0.8462	0.8471	0.8480	32
58	0.8480	0.8490	0.8499	0.8508	0.8517	0.8526	0.8536	0.8545	0.8554	0.8563	0.8572	31
59	0.8572	0.8581	0.8590	0.8599	0.8607	0.8616	0.8625	0.8634	0.8643	0.8652	0.8660	30
60°	0.8660	0.8669	0.8678	0.8686	0.8695	0.8704	0.8712	0.8721	0.8729	0.8738	0.8746	29
61	0.8746	0.8755	0.8763	0.8771	0.8780	0.8788	0.8796	0.8805	0.8813	0.8821	0.8829	28
62	0.8829	0.8838	0.8846	0.8854	0.8862	0.8870	0.8878	0.8886	0.8894	0.8902	0.8910	27
63	0.8910	0.8918	0.8926	0.8934	0.8942	0.8949	0.8957	0.8965	0.8973	0.8980	0.8988	26
64	0.8988	0.8996	0.9003	0.9011	0.9018	0.9026	0.9033	0.9041	0.9048	0.9056	0.9063	25
65	0.9063	0.9070	0.9078	0.9085	0.9092	0.9100	0.9107	0.9114	0.9121	0.9128	0.9135	24
66	0.9135	0.9143	0.9150	0.9157	0.9164	0.9171	0.9178	0.9184	0.9191	0.9198	0.9205	23
67	0.9205	0.9212	0.9219	0.9225	0.9232	0.9239	0.9245	0.9252	0.9259	0.9265	0.9272	22
68	0.9272	0.9278	0.9285	0.9291	0.9298	0.9304	0.9311	0.9317	0.9323	0.9330	0.9336	21
69	0.9336	0.9342	0.9348	0.9354	0.9361	0.9367	0.9373	0.9379	0.9385	0.9391	0.9397	20
70°	0.9397	0.9403	0.9409	0.9415	0.9421	0.9426	0.9432	0.9438	0.9444	0.9449	0.9455	19
71	0.9455	0.9461	0.9466	0.9472	0.9478	0.9483	0.9489	0.9494	0.9500	0.9505	0.9511	18
72	0.9511	0.9516	0.9521	0.9527	0.9532	0.9537	0.9542	0.9548	0.9553	0.9558	0.9563	17
73	0.9563	0.9568	0.9573	0.9578	0.9583	0.9588	0.9593	0.9598	0.9603	0.9608	0.9613	16
74	0.9613	0.9617	0.9622	0.9627	0.9632	0.9636	0.9641	0.9646	0.9650	0.9655	0.9659	15
75	0.9659	0.9664	0.9668	0.9673	0.9677	0.9681	0.9686	0.9690	0.9694	0.9699	0.9703	14
76	0.9703	0.9707	0.9711	0.9715	0.9720	0.9724	0.9728	0.9732	0.9736	0.9740	0.9744	13
77	0.9744	0.9748	0.9751	0.9755	0.9759	0.9763	0.9767	0.9770	0.9774	0.9778	0.9781	12
78	0.9781	0.9785	0.9789	0.9792	0.9796	0.9799	0.9803	0.9806	0.9810	0.9813	0.9816	11
79	0.9816	0.9820	0.9823	0.9826	0.9829	0.9833	0.9836	0.9839	0.9842	0.9845	0.9848	10
80°	0.9848	0.9851	0.9854	0.9857	0.9860	0.9863	0.9866	0.9869	0.9871	0.9874	0.9877	9
81	0.9877	0.9880	0.9882	0.9885	0.9888	0.9890	0.9893	0.9895	0.9898	0.9900	0.9903	8
82	0.9903	0.9905	0.9907	0.9910	0.9912	0.9914	0.9917	0.9919	0.9921	0.9923	0.9925	7
83	0.9925	0.9928	0.9930	0.9932	0.9934	0.9936	0.9938	0.9940	0.9942	0.9943	0.9945	6
84	0.9945	0.9947	0.9949	0.9951	0.9952	0.9954	0.9956	0.9957	0.9959	0.9960	0.9962	5
85	0.9962	0.9963	0.9965	0.9966	0.9968	0.9969	0.9971	0.9972	0.9973	0.9974	0.9976	4
86	0.9976	0.9977	0.9978	0.9979	0.9980	0.9981	0.9982	0.9983	0.9984	0.9985	0.9986	3
87	0.9986	0.9987	0.9988	0.9989	0.9990	0.9990	0.9991	0.9992	0.9993	0.9993	0.9994	2
88	0.9994	0.9995	0.9995	0.9996	0.9996	0.9997	0.9997	0.9997	0.9998	0.9998	0.9998	1
89	0.9998	0.9999	0.9999	0.9999	0.9999	1.0000	1.0000	1.0000	1.0000	1.0000	1.0000	0°
	°1.0	°0.9	°0.8	°0.7	°0.6	°0.5	°0.4	°0.3	°0.2	°0.1	°0.0	Deg.

24. What is the natural sine of a 57.3° angle?
    a. 0.8508
    b. 0.8453
    c. 0.8415
    d. 0.8320

25. What is the natural cosine of a 28.4° angle?
    a. 0.8780
    b. 0.8796
    c. 0.8878
    d. 0.8805

STOP

DIAGNOSTIC READING SURVEY
ANSWER SHEET

NAME _____

PROGRAM _____

DATE _____

VOCABULARY	SPELLING	COMPREHENSION
Sample: _____	Sample: _____	Sample: _____
1. _____	1. _____	1. _____
2. _____	2. _____	2. _____
3. _____	3. _____	3. _____
4. _____	4. _____	4. _____
5. _____	5. _____	5. _____
6. _____	6. _____	6. _____
7. _____	7. _____	7. _____
8. _____	8. _____	8. _____
9. _____	9. _____	9. _____
10. _____	10. _____	10. _____
		11. _____
		12. _____
		13. _____
		14. _____
		15. _____
		16. _____
		17. _____
		18. _____
		19. _____
		20. _____
		21. _____
		22. _____
		23. _____
		24. _____
		25. _____

Do not write in this space!

SCORES:
- Vocabulary: _____
- Spelling: _____
- Comprehension: _____

MAY HAVE DIFFICULTY WITH:

Vocabulary	_____
Spelling	_____
Reference Skills	_____
Table of Contents	_____
Using an Index	_____
Finding Main Idea	_____
Locating Facts	_____
Vocabulary in Context	_____
Following Directions	_____
Reading Charts	_____
Reading Tables	_____

APPENDIX

SECTION I

DENOMINATE NUMBERS

Denominate numbers are numbers that include units of measurement. The units of measurement are arranged from the largest units at the left to the smallest unit at the right.

For example: 6 yd 2 ft 4 in

All basic operations of arithmetic can be performed on denominate numbers.

I. EQUIVALENT MEASURES

Measurements that are equal can be expressed in different terms. For example, 12 in = 1 ft. If these equivalents are divided, the answer is 1.

$$\frac{1 \text{ ft}}{12 \text{ in}} = 1 \qquad \frac{12 \text{ in}}{1 \text{ ft}} = 1$$

To express one measurement as another equal measurement, multiply by the equivalent in the form of 1.

To express 6 inches in equivalent foot measurement, multiply 6 inches by one in the form of $\frac{1 \text{ ft}}{12 \text{ in}}$. In the numerator and denominator, divide by a common factor.

$$6 \text{ in} = \frac{\cancel{6 \text{ in}}^{1}}{1} \times \frac{1}{\cancel{12 \text{ in}}_{2}} = \frac{1}{2} \text{ ft or } 0.5 \text{ ft}$$

To express 4 feet in equivalent inch measurement, multiply 4 feet by one in the form of $\frac{12 \text{ in}}{1 \text{ ft}}$.

$$4 \text{ ft} = \frac{\cancel{4 \text{ ft}}^{4}}{\,} \times \frac{12 \text{ in}}{\cancel{1 \text{ ft}}_{1}} = \frac{48 \text{ in}}{1} = 48 \text{ in}$$

Per means division, as with a fraction bar. For example, 50 miles per hour can be written $\frac{50 \text{ miles}}{1 \text{ hour}}$.

II. BASIC OPERATIONS

A. ADDITION

SAMPLE: 2 yd 1 ft 5 in + 1 ft 8 in + 5 yd 2 ft

1. Write the denominate numbers in a column with like units in the same column.

2. Add the denominate numbers in each column.

3. Express the answer using the largest possible units.

```
        2 yd   1 ft    5 in
               1 ft    8 in
    +   5 yd   2 ft
        ─────────────────────
        7 yd   4 ft   13 in

        7 yd                  =  7 yd
               4 ft           =  1 yd   1 ft
                     13 in  = +           1 ft   1 in
                              ──────────────────────────
        7 yd   4 ft   13 in  =  8 yd    2 ft   1 in
```

B. SUBTRACTION

SAMPLE: 4 yd 3 ft 5 in - 2 yd 1 ft 7 in

1. Write the denominate numbers in columns with like units in the same column.

2. Starting at the right, examine each column to compare the numbers. If the bottom number is larger, exchange one unit from the column at the left for its equivalent. Combine like units.

3. Subtract the denominate numbers.

4. Express the answer using the largest possible units.

```
        4 yd   3 ft    5 in
    -   2 yd   1 ft    7 in
```

7 in is larger than 5 in

3 ft = 2 ft 12 in
12 in + 5 in = 17 in

```
        4 yd   2 ft   17 in
    -   2 yd   1 ft    7 in
        ─────────────────────
        2 yd   1 ft   10 in

        2 yd   1 ft   10 in
```

C. MULTIPLICATION

—By a constant

SAMPLE: 1 hr 24 min X 3

1. Multiply the denominate number by the constant.

2. Express the answer using the largest possible units.

$$\begin{array}{r} 1\text{ hr }\ 24\text{ min} \\ \times\ 3 \\ \hline 3\text{ hr }\ 72\text{ min} \end{array}$$

3 hr = 3 hr
72 min = 1 hr 12 min
3 hr 72 min = 4 hr 12 min

— By a denominate number expressing linear measurement

SAMPLE: 9 ft 6 in X 10 ft

1. Express all denominate numbers in the same unit.

2. Multiply the denominate numbers. (This includes the units of measure, such as ft X ft = sq ft.)

$9\text{ ft }6\text{ in} = 9\frac{1}{2}\text{ ft}$

$9\frac{1}{2}\text{ ft} \times 10\text{ ft} =$

$\frac{19}{2}\text{ ft} \times 10\text{ ft} =$

95 sq ft

— By a denominate number expressing square measurement

SAMPLE: 3 ft X 6 sq ft

1. Multiply the denominate numbers. (This includes the units of measure, such as ft X ft = sq ft and sq ft X ft = cu ft.)

3 ft X 6 sq ft = 18 cu ft

— By a denominate number expressing rate

SAMPLE: 50 miles per hour X 3 hours

1. Express the rate as a fraction using the fraction bar for *per*.

2. Divide the numerator and denominator by any common factors, including units of measure.

$\frac{50\text{ miles}}{1\text{ hour}} \times \frac{3\text{ hours}}{1} =$

$\frac{50\text{ miles}}{\cancel{1\text{ hour}}} \times \frac{\cancel{3\text{ hours}}^{3}}{1} =$

3. Multiply numerators. $\dfrac{150 \text{ miles}}{1} =$
   Multiply denominators.

4. Express the answer in the remaining unit.   150 miles

D. **DIVISION**

   — By a constant

   SAMPLE: 8 gal 3 qt ÷ 5

   1. Express all denominate numbers in the same unit.   8 gal 3 qt = 35 qt

   2. Divide the denominate number by the constant.   35 qt ÷ 5 = 7 qt

   3. Express the answer using the largest possible units.   7 qt = 1 gal 3 qt

   — By a denominate number expressing linear measurement

   SAMPLE: 11 ft 4 in ÷ 8 in

   1. Express all denominate numbers in the same unit.   11 ft 4 in = 136 in

   2. Divide the denominate numbers by a common factor. (This includes the units of measure, such as inches ÷ inches = 1.)

   $136 \text{ in} \div 8 \text{ in} = 17$

   $\dfrac{\cancel{136 \text{ in}}}{\cancel{8 \text{ in}}} = \dfrac{17}{1} = 17$

   — By a linear measure with a square measurement as the dividend

   SAMPLE: 20 sq ft ÷ 4 ft

   1. Divide the denominate numbers. (This includes the units of measure, such as sq ft ÷ ft = ft.)

   20 sq ft ÷ 4 ft
   5 ft

   $\dfrac{\cancel{20 \text{ sq ft}}}{\cancel{4 \text{ ft}}} = \dfrac{5 \text{ ft}}{1}$

   2. Express the answer in the remaining unit.   5 ft

— *By denominate numbers used to find rate*

SAMPLE: 200 mi ÷ 10 gal

1. Divide the denominate numbers.

$$\frac{200 \text{ mi}}{10 \text{ gal}} = \frac{20 \text{ mi}}{1 \text{ gal}}$$

2. Express the units with the fraction bar meaning *per*.

$$\frac{20 \text{ mi}}{1 \text{ gal}} = 20 \text{ miles per gallon}$$

Note: Alternate methods of performing the basic operations will produce the same result. The choice of method is determined by the individual.

SECTION II

EQUIVALENTS

ENGLISH RELATIONSHIPS

ENGLISH LENGTH MEASURE

1 foot (ft) = 12 inches (in)
1 yard (yd) = 3 feet (ft)
1 mile (mi) = 1 760 yards (yd)
1 mile (mi) = 5 280 feet (ft)

ENGLISH AREA MEASURE

1 square yard (sq yd) = 9 square feet (sq ft)
1 square foot (sq ft) = 144 square inches (sq in)
1 square mile (sq mi) = 640 acres
1 acre = 43 560 square feet (sq ft)

ENGLISH VOLUME MEASURE FOR SOLIDS

1 cubic yard (cu yd) = 27 cubic feet (cu ft)
1 cubic foot (cu ft) = 1 728 cubic inches (cu in)

ENGLISH VOLUME MEASURE FOR FLUIDS

1 quart (qt) = 2 pints (pt)
1 gallon (gal) = 4 quarts (qt)

ENGLISH VOLUME MEASURE EQUIVALENTS

1 gallon (gal) = 0.133 681 cubic foot (cu ft)
1 gallon (gal) = 231 cubic inches (cu in)

SI METRICS STYLE GUIDE

SI metrics is derived from the French name Le Système International d'Unités. The metric unit names are already in accepted practice. SI metrics attempts to standardize the names and usages so that students of metrics will have a universal knowledge of the application of terms, symbols, and units.

The English system of mathematics (used in the United States) has always had many units in its weights and measures tables which were not applied to everyday use. For example, the pole, perch, furlong, peck, and scruple are not used often. These measurements, however, are used to form other measurements and it has been necessary to include the measurements in the tables. Including these measurements aids in the understanding of the orderly sequence of measurements greater or smaller than the less frequently used units.

The metric system also has units that are not used in everyday application. Only by learning the lesser-used units is it possible to understand the order of the metric system. SI metrics, however, places an emphasis on the most frequently used units.

In using the metric system and writing its symbols, certain guidelines are followed. For the student's reference, some of the guidelines are listed.

1. In using the symbols for metric units, the first letter is capitalized only if it is derived from the name of a person.

   SAMPLE:

UNIT	SYMBOL	UNIT	SYMBOL
metre	m	Newton	N (named after Sir Isaac Newton)
gram	g	degree Celsius	°C (named after Anders Celsius)

   EXCEPTION: The symbol for litre is L. This is used to distinguish it from the number one (1).

2. Prefixes are written with lowercase letters.

   SAMPLE:

PREFIX	UNIT	SYMBOL
centi	metre	cm
milli	gram	mg

   EXCEPTIONS:

PREFIX	UNIT	SYMBOL
tera	metre	Tm (used to distinguish it from the metric tonne, t)
giga	metre	Gm (used to distinguish it from gram, g)
mega	gram	Mg (used to distinguish it from milli, m)

3. Periods are not used in the symbols. Symbols for units are the same in the singular and the plural (no "s" is added to indicate a plural).

   SAMPLE: 1 mm *not* 1 mm.   3 mm *not* 3 mms

4. When referring to a unit of measurement, symbols are not used. The symbol is used only when a number is associated with it.

   SAMPLE: The length of the room is expressed in metres.   *not*   The length of the room is expressed in m. (*The length of the room is 25 m is correct.*)

5. When writing measurements that are less than one, a zero is written before the decimal point.

   SAMPLE: 0.25 m *not* .25 m

6. Separate the digits in groups of three, counting from the decimal point to the left and to the right. A space is left between the groups of digits.

   SAMPLE: 5 179 232 mm *not* 5,179,232 mm   0.566 23 mg *not* 0.56623 mg   1 346.098 7 L *not* 1,346.0987 L

A space is also left between the digits and the unit of measure.
SAMPLE: 5 179 232 mm *not* 5 179 232mm

7. Symbols for area measure and volume measure are written with exponents.
   SAMPLE: 3 cm^2 *not* 3 sq. cm   4 km^3 *not* 4 cu. km

8. Metric words with prefixes are accented on the first syllable. In particular, kilometre is pronounced "kill'-o-metre." This avoids confusion with words for measuring devices which are generally accented on the second syllable, such as thermometer (ther-mom'-e-ter).

METRIC RELATIONSHIPS

The base units in SI metrics include the metre and the gram. Other units of measure are related to these units. The relationship between the units is based on powers of ten and uses these prefixes:

    kilo (1 000)   hecto (100)   deka (10)   deci (0.1)   centi (0.01)   milli (0.001)

These tables show the most frequently used units with an asterisk (*).

METRIC LENGTH MEASURE

10 millimetres (mm)*	=	1 centimetre (cm)*
10 centimetres (cm)	=	1 decimetre (dm)
10 decimetres (dm)	=	1 metre (m)*
10 metres (m)	=	1 dekametre (dam)
10 dekametres (dam)	=	1 hectometre (hm)
10 hectometres (hm)	=	1 kilometre (km)*

▲ To express a metric length unit as a smaller metric length unit, multiply by a positive power of ten such as 10, 100, 1 000, 10 000, etc.

▲ To express a metric length unit as a larger metric length unit, multiply by a negative power of ten such as 0.1, 0.01, 0.001, 0.0001, etc.

METRIC AREA MEASURE

100 square millimetres (mm^2)	=	1 square centimetre (cm^2)*
100 square centimetres (cm^2)	=	1 square decimetre (dm^2)
100 square decimetres (dm^2)	=	1 square metre (m^2)*
100 square metres (m^2)	=	1 square dekametre (dam^2)
100 square dekametres (dam^2)	=	1 square hectometre (hm^2)*
100 square hectometres (hm^2)	=	1 square kilometre (km^2)

▲ To express a metric area unit as a smaller metric area unit, multiply by 100, 10 000, 1 000 000, etc.

▲ To express a metric area unit as a larger metric area unit, multiply by 0.01, 0.000 1, 0.000 001, etc.

METRIC VOLUME MEASURE FOR SOLIDS

1 000 cubic millimetres (mm^3)	=	1 cubic centimetre (cm^3)*
1 000 cubic centimetres (cm^3)	=	1 cubic decimetre (dm^3)*
1 000 cubic decimetres (dm^3)	=	1 cubic metre (m^3)*
1 000 cubic metres (m^3)	=	1 cubic dekametre (dam^3)
1 000 cubic dekametres (dam^3)	=	1 cubic hectometre (hm^3)
1 000 cubic hectometres (hm^3)	=	1 cubic kilometre (km^3)

▲ To express a metric volume unit for solids as a smaller metric volume unit for solids, multiply by 1 000, 1 000 000, 1 000 000 000, etc.

▲ To express a metric volume unit for solids as a larger metric volume unit for solids, multiply by 0.001, 0.000 001, 0.000 000 001, etc.

METRIC VOLUME MEASURE FOR FLUIDS

10 millilitres (mL)*	=	1 centilitre (cL)
10 centilitres (cL)	=	1 decilitre (dL)
10 decilitres (dL)	=	1 litre (L)*
10 litres (L)	=	1 dekalitre (daL)
10 dekalitres (daL)	=	1 hectolitre (hL)
10 hectolitres (hL)	=	1 kilolitre (kL)

▲ To express a metric volume unit for fluids as a smaller metric volume unit for fluids, multiply by 10, 100, 1 000, 10 000, etc.

▲ To express a metric volume unit for fluids as a larger metric volume unit for fluids, multiply by 0.1, 0.01, 0.001, 0.000 1, etc.

METRIC VOLUME MEASURE EQUIVALENTS

1 cubic decimetre (dm^3)	=	1 litre (L)
1 000 cubic centimetres (cm^3)	=	1 litre (L)
1 cubic centimetre (cm^3)	=	1 millilitre (mL)

METRIC MASS MEASURE

10 milligrams (mg)*	=	1 centigram (cg)
10 centigrams (cg)	=	1 decigram (dg)
10 decigrams (dg)	=	1 gram (g)*
10 grams (g)	=	1 dekagram (dag)
10 dekagrams (dag)	=	1 hectogram (hg)
10 hectograms (hg)	=	1 kilogram (kg)*
1 000 kilograms (kg)	=	1 megagram (Mg)*

▲ To express a metric mass unit as a smaller metric mass unit, multiply by 10, 100, 1 000, 10 000, etc.

▲ To express a metric mass unit as a larger metric mass unit, multiply by 0.1, 0.01, 0.001, 0.000 1, etc.

Metric measurements are expressed in decimal parts of a whole number. For example, one-half millimetre is written as 0.5 mm.

In calculating with the metric system, all measurements are expressed using the same prefixes. If answers are needed in millimetres, all parts of the problem should be expressed in millimetres before the final solution is attempted. Diagrams that have dimensions in different prefixes must first be expressed using the same unit.

ENGLISH-METRIC EQUIVALENTS

LENGTH MEASURE

1 inch (in)	=	25.4 millimetres (mm)
1 inch (in)	=	2.54 centimetres (cm)
1 foot (ft)	=	0.304 8 metre (m)
1 yard (yd)	=	0.914 4 metre (m)
1 mile (mi)	≈	1.609 kilometres (km)
1 millimetre (mm)	≈	0.039 37 inch (in)
1 centimetre (cm)	≈	0.393 70 inch (in)
1 metre (m)	≈	3.280 84 feet (ft)
1 metre (m)	≈	1.093 61 yards (yd)
1 kilometre (km)	≈	0.621 37 mile (mi)

MASS MEASURE

1 pound (lb)	≈	0.453 592 kilogram (kg)
1 pound (lb)	≈	453.592 37 grams (g)
1 ounce (oz)	≈	28.349 523 grams (g)
1 ounce (oz)	≈	0.028 350 kilogram (kg)
1 kilogram (kg)	≈	2.204 623 pounds (lb)
1 gram (g)	≈	0.002 205 pound (lb)
1 kilogram (kg)	≈	35.273 962 ounces (oz)
1 gram (g)	≈	0.035 274 ounce (oz)

AREA MEASURE

1 square inch (sq in)	=	645.16 square millimetres (mm^2)
1 square inch (sq in)	=	6.451 6 square centimetres (cm^2)
1 square foot (sq ft)	≈	0.092 903 square metre (m^2)
1 square yard (sq yd)	≈	0.836 127 square metre (m^2)
1 square millimetre (mm^2)	≈	0.001 550 square inch (sq in)
1 square centimetre (cm^2)	≈	0.155 00 square inch (sq in)
1 square metre (m^2)	≈	10.763 910 square feet (sq ft)
1 square metre (m^2)	≈	1.119 599 square yards (sq yd)

VOLUME MEASURE FOR SOLIDS

1 cubic inch (cu in)	=	16.387 064 cubic centimetres (cm^3)
1 cubic foot (cu ft)	≈	0.028 317 cubic metre (m^3)
1 cubic yard (cu yd)	≈	0.764 555 cubic metre (m^3)
1 cubic centimetre (cm^3)	≈	0.061 024 cubic inch (cu in)
1 cubic metre (m^3)	≈	35.314 667 cubic feet (cu ft)
1 cubic metre (m^3)	≈	1.307 951 cubic yards (cu yd)

VOLUME MEASURE FOR FLUIDS

1 gallon (gal)	≈	3 785.411 cubic centimetres (cm^3)
1 gallon (gal)	≈	3.785 411 litres (L)
1 quart (qt)	≈	0.946 353 litre (L)
1 ounce (oz)	≈	29.573 530 cubic centimetres (cm^3)
1 cubic centimetre (cm^3)	≈	0.000 264 gallon (gal)
1 litre (L)	≈	0.264 172 gallon (gal)
1 litre (L)	≈	1.056 688 quarts (qt)
1 cubic centimetre (cm^3)	≈	0.033 814 ounce (oz)

SECTION III

FORMULAS

CIRCULAR MIL FORMULA

$$A = d^2$$

where A = area in circular mils
d = diameter in mils

TEMPERATURE

$$°C = \frac{5}{9}(°F - 32°)$$

PYTHAGOREAN THEOREM

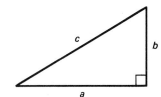

$$c^2 = a^2 + b^2$$
$$a^2 = c^2 - b^2$$
$$b^2 = c^2 - a^2$$

where c = the hypotenuse
a = the side of the triangle
b = the side of the triangle

TRIGONOMETRIC FORMULAS

$$\text{sine of an angle} = \frac{\text{opposite side}}{\text{hypotenuse}}$$

$$\text{cosine of an angle} = \frac{\text{adjacent side}}{\text{hypotenuse}}$$

$$\text{tangent of an angle} = \frac{\text{opposite side}}{\text{adjacent side}}$$

VOLTAGE AND POWER FORMULAS

$E = IR$	$P = EI$
$E = \frac{P}{I}$	$P = \frac{E^2}{R}$
$E = \sqrt{PR}$	$P = I^2 R$
$R = \frac{E}{I}$	$I = \frac{E}{R}$
$R = \frac{E^2}{P}$	$I = \frac{P}{E}$
$R = \frac{P}{I^2}$	$I = \sqrt{\frac{P}{R}}$

where E = voltage in volts
R = resistance in ohms
I = current in amperes
P = power in watts

IMPEDANCE FORMULA

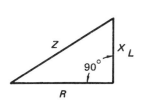

IMPEDANCE DIAGRAM

$$Z^2 = R^2 + X_L^2$$

$$Z = \sqrt{R^2 + X_L^2}$$

$$X_L = \sqrt{Z^2 - R^2}$$

$$R = \sqrt{Z^2 - X_L^2}$$

where Z = impedance
X_L = reactance
R = resistance

VOLTAGE FORMULA

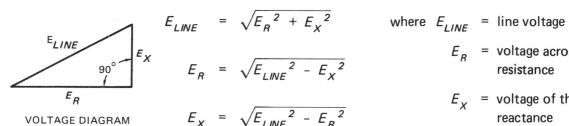

VOLTAGE DIAGRAM

$$E_{LINE} = \sqrt{E_R^2 + E_X^2}$$

$$E_R = \sqrt{E_{LINE}^2 - E_X^2}$$

$$E_X = \sqrt{E_{LINE}^2 - E_R^2}$$

where E_{LINE} = line voltage
E_R = voltage across the resistance
E_X = voltage of the reactance

SECTION IV

TRIGONOMETRIC FUNCTIONS

Angle	Sine	Cosine	Tangent	Angle	Sine	Cosine	Tangent
1°	0.017 5	0.999 8	0.017 5	46°	0.719 3	0.694 7	1.035 5
2°	0.034 9	0.999 4	0.034 9	47°	0.731 4	0.682 0	1.072 4
3°	0.052 3	0.998 6	0.052 4	48°	0.743 1	0.669 1	1.110 6
4°	0.069 8	0.997 6	0.069 9	49°	0.754 7	0.656 1	1.150 4
5°	0.087 2	0.996 2	0.087 5	50°	0.766 0	0.642 8	1.191 8
6°	0.104 5	0.994 5	0.105 1	51°	0.777 1	0.629 3	1.234 9
7°	0.121 9	0.992 5	0.122 8	52°	0.788 0	0.615 7	1.279 9
8°	0.139 2	0.990 3	0.140 5	53°	0.798 6	0.601 8	1.327 0
9°	0.156 4	0.987 7	0.158 4	54°	0.809 0	0.587 8	1.376 4
10°	0.173 6	0.984 8	0.176 3	55°	0.819 2	0.573 6	1.428 1
11°	0.190 8	0.981 6	0.194 4	56°	0.829 0	0.559 2	1.482 6
12°	0.207 9	0.978 1	0.212 6	57°	0.838 7	0.544 6	1.539 9
13°	0.225 0	0.974 4	0.230 9	58°	0.848 0	0.529 9	1.600 3
14°	0.241 9	0.970 3	0.249 3	59°	0.857 2	0.515 0	1.664 3
15°	0.258 8	0.965 9	0.267 9	60°	0.866 0	0.500 0	1.732 1
16°	0.275 6	0.961 3	0.286 7	61°	0.874 6	0.484 8	1.804 0
17°	0.292 4	0.956 3	0.305 7	62°	0.882 9	0.469 5	1.880 7
18°	0.309 0	0.951 1	0.324 9	63°	0.891 0	0.454 0	1.962 6
19°	0.325 6	0.945 5	0.344 3	64°	0.898 8	0.438 4	2.050 3
20°	0.342 0	0.939 7	0.364 0	65°	0.906 3	0.422 6	2.144 5
21°	0.358 4	0.933 6	0.383 9	66°	0.913 5	0.406 7	2.246 0
22°	0.374 6	0.927 2	0.404 0	67°	0.920 5	0.390 7	2.355 9
23°	0.390 7	0.920 5	0.424 5	68°	0.927 2	0.374 6	2.475 1
24°	0.406 7	0.913 5	0.445 2	69°	0.933 6	0.358 4	2.605 1
25°	0.422 6	0.906 3	0.466 3	70°	0.939 7	0.342 0	2.747 5
26°	0.438 4	0.898 8	0.487 7	71°	0.945 5	0.325 6	2.904 2
27°	0.454 0	0.891 0	0.509 5	72°	0.951 1	0.309 0	3.077 7
28°	0.469 5	0.882 9	0.531 7	73°	0.956 3	0.292 4	3.270 9
29°	0.484 8	0.874 6	0.554 3	74°	0.961 3	0.275 6	3.487 4
30°	0.500 0	0.866 0	0.577 4	75°	0.965 9	0.258 8	3.732 1
31°	0.515 0	0.857 2	0.600 9	76°	0.970 3	0.241 9	4.010 8
32°	0.529 9	0.848 0	0.624 9	77°	0.974 4	0.225 0	4.331 5
33°	0.544 6	0.838 7	0.649 4	78°	0.978 1	0.207 9	4.704 6
34°	0.559 2	0.829 0	0.674 5	79°	0.981 6	0.190 8	5.144 6
35°	0.573 6	0.819 2	0.700 2	80°	0.984 8	0.173 6	5.671 3
36°	0.587 8	0.809 0	0.726 5	81°	0.987 7	0.156 4	6.313 8
37°	0.601 8	0.798 6	0.753 6	82°	0.990 3	0.139 2	7.115 4
38°	0.615 7	0.788 0	0.781 3	83°	0.992 5	0.121 9	8.144 3
39°	0.629 3	0.777 1	0.809 8	84°	0.994 5	0.104 5	9.514 4
40°	0.642 8	0.766 0	0.839 1	85°	0.996 2	0.087 2	11.430 1
41°	0.656 1	0.754 7	0.869 3	86°	0.997 6	0.069 8	14.300 7
42°	0.669 1	0.743 1	0.900 4	87°	0.998 6	0.052 3	19.081 1
43°	0.682 0	0.731 4	0.932 5	88°	0.999 4	0.034 9	28.636 3
44°	0.694 7	0.719 3	0.965 7	89°	0.999 8	0.017 5	57.290 0
45°	0.707 1	0.707 1	1.000 0	90°	1.000 0	0.000 0	

GLOSSARY

Alternating current — (AC) An electric current that reverses direction at regularly recurring intervals.

American wire gauge — (AWG) A system of numbers used in sizing wire.

Ammeter — An instrument used for measuring electric current in amperes.

Ampere — (A) A unit of electric current.

Askarel — A synthetic, noncombustible, insulating liquid used in transformers.

Ballast — A resistance used in fluorescent fixtures to stabilize the current.

Bus bar — A heavy copper strip or bar used as a primary power source to carry heavy currents or to make a common connection between several circuits.

Bushing — An insulating sleeve inserted in an opening in a metal plate to protect a through conductor.

Cabinet — A flush or surface mounted box used as a protective housing for electrical equipment.

Cable — A stranded assembly of one or more conductors, usually within a protective sheath.

Capacitance — (C) The property of a conductor or capacitor that permits the storage of electrical energy. The unit of capacitance is the farad.

Capacitive reactance — (X_c) The opposition to alternating current due to capacitance. The capacitive reactance is expressed in ohms.

Capacitor — An electrical device consisting of two conducting surfaces or sets of conducting surfaces which are oppositely charged and are separated by a thin layer of insulating material such as air, paraffin paper, or mica.

Circuit — The conducting part or a system of conducting parts through which an electric current is intended to flow.

Circuit breaker — A device which automatically opens a circuit if the current exceeds the rated amount. The circuit breaker does not destroy itself as a fuse does.

Circular mil — The cross-sectional area of a wire which is 1 mil (0.001 inch) in diameter.

Coil — An electrical element in the circuit consisting of a spiral of wire either self-supporting or wound on a spool or other structure usually for electromagnetic effect or for providing resistance.

Conductance — (G) The ability of material to carry electrical current. It is measured in siemens.

Conductor — A substance capable of transmitting electricity.

Conduit — A pipe or tube used for receiving and protecting electric wires and cables.

Coulomb — (C) A unit of electric charge equal to 6.25×10^{18} electrons.

Current — (I) The transfer of electric charge through a conductor. Current is measured in amperes.

Cycle — A complete positive and a complete negative alternation of voltage or current.

Device — An item in an electrical system for the purpose of carrying but not using electricity.

Dielectric — A nonconductor used between the plates of a capacitor.

Dielectric constant — (k) A measure of the ability of a dielectric material to store electrical potential energy.

Direct current — (DC) An electric current in which there is a continuous transfer of charge in only one direction.

Efficiency — (Eff) A percent value expressing the ratio of power output to power input.

Electrical Metallic Tubing — (EMT) A thin walled circular metal raceway used for pulling in or withdrawing wires or cables.

Electrolyte — A nonmetallic electric conductor in which current is carried by the movement of ions used in wet cell batteries.

Farad — (F) The unit of measure for capacitance.

Feeder — A heavy wire conductor connecting points of an electric distribution system such as a substation and a generating station.

Fitting — A small accessory, generally standardized, used for a mechanical rather than electrical function as a locknut or bushing.

Frequency — (f) The number of complete cycles in a unit of time. It is expressed in hertz.

Fuse — An electrical safety device of wire or a strip of fusible metal that melts and opens the electrical circuit when the current exceeds the rated amount. It destroys itself and must be replaced.

Generator — A machine that changes mechanical energy into electrical energy of measure.

Hertz — (Hz) The unit of measure for frequency.

Henry — (H) The unit of measure of inductance.

Horsepower — (hp) A unit of power equal to 776 watts of electrical power.

Impedance — (Z) The total opposition to alternating current including inductive reactance, capacitive reactance, and resistance. It is expressed in ohms.

Inductance — (L) The property of an alternating current circuit to induce an electromotive force by variations of current. It is expressed in henrys.

Induction — The process of producing electrification, magnetization, or induced voltage in an object by exposure to a magnetic field or a charged body.

Inductive reactance — (X_L) The opposition to an alternating current. It is expressed in ohms.

Inductor — A conductor that acts upon another or is itself acted upon by induction. As the conductor is wound into a spiral or coil the inductive intensity increases.

Insulator — A material which has a high resistance to the flow of an electric current.

Internal resistance — The resistance within the source of the electromotive force.

Joule — (J) A unit of electrical energy.

Junction box — A box for inserting and joining cables or wires.

Kilowatt — (kW) A unit of electrical power equal to 1 000 watts.

Kirchhoff's Law — The sum of the currents entering a junction is equal to the sum of the currents leaving that junction.

Line voltage — (V_{LINE}) The voltage existing at the wall outlets or terminals of a power line.

Live — A term which means electrically connected to a source of potential difference or electrically charged to have a potential different from that of the earth. Other terms used for live are alive, hot, or energized.

Locknut — A nut constructed to lock itself when screwed up tight.

Mutual inductance — (M) The condition that exists in a circuit when the positions of two inductors cause magnetic lines of force from one inductor to link the turns of the other.

Negative — (-) A terminal or electrode with excess electron.

Ohm — (Ω) A unit of measure resistance.

Ohm's Law — Current is directly proportional to voltage and inversely proportional to resistance.

Oscillator — A device for producing alternating current power at a frequency determined by the values of certain constants in the circuit.

Outlet — A set of mounted and insulated terminals to which electric appliances may be connected such as a receptacle or an electric socket.

Parallel circuit — A method of connecting a circuit so the current has two or more paths to follow.

Positive — (+) A terminal or electrode with a deficiency of electrons.

Potential difference — The difference in electrical pressures between two points in a circuit. It is measured in volts.

Power — (P) The rate of doing work and expressed in watts.

Power factor — (PF) The rate of actual power in an alternating current circuit compared to the mathematically determined power. It can be expressed as a decimal or as a percent.

Primary voltage — (E_p) The voltage of the circuit supplying power to a transformer. It is the input voltage.

Raceway — A channel for loosely holding electrical wires.

Reactance — (X) Opposition to alternating current due to inductance and capacitance. It is expressed in ohms.

Receptacle — A mounted female electrical fitting that contains the live parts of the circuit.

Resistance — (R) The opposition a material offers to the flow of electrons. It is expressed in ohms.

Resistor — A device that opposes the flow of an electric current and is used for protection, operation, or current control.

Secondary voltage — (E_s) The voltage output of a transformer.

Series circuit — A method of connecting a circuit so the current has one path to follow.

Shunt — A conductor connected in parallel with another component in an electric circuit.

Solder — An alloy of lead and tin which melts at a low temperature and is used to join metallic surfaces.

Solenoid — An electromagnetic coil with a moveable core which is drawn into the coil when the current flows.

Switch — A mechanical device for completing, interrupting, or changing the connections in an electrical circuit.

Switchboard — A type of switch-gear assembly which consists of one or more panels with mounted electrical devices.

Transformer — An electromagnetic device using induction to increase or decrease alternating current voltage.

Transistor — An electronic device consisting of a small block of semiconductors with at least three electrodes.

Volt — (V) A unit of electrical potential or pressure.

Voltage — (E) The electromotive force or electrical pressure. It is expressed in volts.

Watt — (W) A unit of measure of power.

Wavelength — The distance traveled by a wave during the time interval covered by a cycle.

Wheatstone bridge — An apparatus which is used to measure resistance by varying known resistances until the system is balanced.

ANSWERS TO ODD-NUMBERED PROBLEMS

SECTION 1 WHOLE NUMBERS

UNIT 1 ADDITION OF WHOLE NUMBERS

1. 1 165
3. 556
5. $378
7. 2 380 m
9. 615 lb
11. 14 839 W
13. 473
15. 1 052
17. 3 706 cm

UNIT 2 SUBTRACTION OF WHOLE NUMBERS

1. 1 161 ft
3. 455
5. 9 m
7. 305 ft
9. 200
11. 1 372 kW · h
13. 65
15. 221 ft
17. 10 MΩ

UNIT 3 MULTIPLICATION OF WHOLE NUMBERS

1. a. 48
   b. 63
   c. 33
   d. 32
   e. 42
   f. 22
3. 16 412 W
5. No
7. 1 980 W
9. $8 800
11. 1 940 W
13. 220 circuits

UNIT 4 DIVISION OF WHOLE NUMBERS

1. 47 staples
3. 5 124 W
5. 5 outlets
7. 21 ft
9. 8 weeks
11. 36 lamps
13. $1
15. 1 320 W
17. 30 m

UNIT 5 COMBINED OPERATIONS WITH WHOLE NUMBERS

1. 591
3. 698 lb
5. 1 192 ft
7. 652 ft
9. $95
11. 11 250 W
13. 392
15. 91
17. 4 000 ft

SECTION 2 COMMON FRACTIONS

UNIT 6 ADDITION OF COMMON FRACTIONS

1. a. 1 21/32
   b. 2 21/32
   c. 2 17/48
   d. 2 11/20
3. 1 7/8 hp
5. 3/16 in
7. 1 7/64 in
9. 7/16 in
11. 6 5/16 in
13. 1 1/64 in
15. 19 1/4 in
17. 43 9/10 m

UNIT 7 SUBTRACTION OF COMMON FRACTIONS

1. 3/4 in
3. 51/64 in
5. 3/16 in
7. 863/1 000 in
9. 376/1 000 in
11. 3 1/32 in
13. 1 7/64 in
15. 153 7/10 m

UNIT 8 MULTIPLICATION OF COMMON FRACTIONS

1. 7 9/16 lb
3. 255 7/12 ft
5. 264 7/16 ft
7. 26 9/16 kW · h
9. 5/12 kg

UNIT 9 DIVISION OF COMMON FRACTIONS

1. 8
3. 1 41/49 W
5. 35
7. 1 13/17
9. $4

UNIT 10 COMBINED OPERATIONS WITH COMMON FRACTIONS

1. 2 41/48 hp
3. 21/32 in
5. 7/8 in
7. 26 11/16 in
9. 1/8 in
11. 469/1 000 in
13. 251 3/4'
15. 44 1/2 ft
17. $400
19. $3
21. 1 193 41/60
23. 30
25. 5 000 ft
27. 49

SECTION 3 DECIMAL FRACTIONS

UNIT 11 ADDITION OF DECIMAL FRACTIONS

1. $112.16
3. $11.10
5. $22.94
7. $22.89
9. 9.724 A
11. 2.312 5 in
13. 7 in
15. 2.687 5 in
17. 4.635 cm

UNIT 12 SUBTRACTION OF DECIMAL FRACTIONS

1. 1 067.35 kW · h
3. 9.12 lb
5. 0.625 5 in
7. 3.375 in
9. 1.25 in
11. 1.062 5 in
13. 1.062 5 in
15. larger
17. #12
19. 0.003 5 in

UNIT 13 MULTIPLICATION OF DECIMAL FRACTIONS

1. $0.063
3. 58.12 in
5. 7.461 in
7. 15.361 5 lb
9. $27.15
11. $433.13
13. $2 084.58
15. 31.416 ft/min
17. 37.125 V

Answers to Odd-Numbered Problems 183

UNIT 14 DIVISION OF DECIMAL FRACTIONS

1. $1.066
3. 0.002 56 Ω
5. $0.86
7. 2.406 25 in
9. 180.82 W
11. $188.58
13. 1 676.75 lb

UNIT 15 DECIMAL AND COMMON FRACTION EQUIVALENTS

1. 1.312 5 in
3. 25.39 lb
5. 0.131 in
7. 1 5/8 in
9. 1/8 in
11. 1.753 in
13. 0.258 in
15. 0.432 5 in by 0.622 in

UNIT 16 COMBINED OPERATIONS WITH DECIMAL FRACTIONS

1. 0.535 in
3. 2.171 A
5. 2.217 5 in
7. 3.86 in
9. 3.555 in
11. 8.375 lb
13. 1.5 A
15. $40.24
17. $8.50
19. 7.04 in

SECTION 4 PERCENTS, AVERAGES, AND ESTIMATES

UNIT 17 PERCENT AND PERCENTAGE

1. 20%
3. $39.74
5. 592
7. $28.71
9. 90%
11. $66.50

UNIT 18 INTEREST

1. $1 071
3. $486
5. $4 290.33
7. $2 146.71
9. $373.75
11. $234

UNIT 19 DISCOUNT

1. $31.60
3. $752.06
5. $74.64
7. $981
9. $1 103.42
11. $1 425.60

UNIT 20 AVERAGES AND ESTIMATES

1. 51.76 kW · h
3. $75.25
5. a. $18.21
   b. $3.94 over
7. a. 9.5 coils
   b. 36.1 m over
9. 35.7° F

UNIT 21 COMBINED PROBLEMS ON PERCENTS, AVERAGES, AND ESTIMATES

1. 16.7%
3. 82.8 hp
5. $177
7. $105
9. $265.22
11. $1 745.17
13. 94%

SECTION 5 POWERS AND ROOTS

UNIT 22 POWERS

1. 49
3. 81
5. 121
7. 512
9. 10 000
11. 225
13. 15 625
15. 810 000
17. 147 008 443
19. 4 096 CM
21. 1 024 CM
23. 201.6 W
25. 13 444.4 W
27. 4 215.687 7 CM

UNIT 23 SQUARE ROOTS

1. 9
3. 13
5. 23
7. 29.87
9. 41.13
11. 32 A
13. 110 V
15. 2 A
17. 3.95 A
19. 0.21 A
21. 2.50 A
23. 101.88 mils
25. 127.28 V
27. 3.64 A
29. 110.88 mils

UNIT 24 COMBINED OPERATIONS WITH POWERS AND ROOTS

1. 11
3. 17
5. 27
7. 39
9. 56
11. 10 383.61 CM
13. 81 mils
15. 2.77 A
17. 3.74 A

SECTION 6 MEASURE

UNIT 25 LENGTH MEASURE

1. 1 200 mm
3. 37 mm
5. 0.43 m
7. 15.24 cm
9. 88.495 km
11. 3 lengths
13. 2.85 mm
15. 3.09 m
17. a. 8.22 cm
    b. 13.3 cm

UNIT 26 AREA MEASURE

1. 750 000 mm^2
3. 2.5 sq ft
5. 2.51 m^2
7. 101.96 cm^2
9. 6 fixtures
11. 30.38 m^2
13. 61 m
15. 47.694 cm^2
17. a. 6 000 W
    b. 50 A
    c. 4 circuits

UNIT 27 VOLUME AND MASS MEASURE

1. 1 200 000 000 mm^3
3. 1.56 cu yd
5. 5.68 L
7. 6.88 m^3
9. 0.142 m^3
11. 9 818 gal
13. 31.7 m^3

Answers to Odd-Numbered Problems 185

UNIT 28 ENERGY AND TEMPERATURE MEASURE

1. a. 1.5×10^3 mA
   b. 1.5×10^6 μA
3. a. 2×10^{-6} F
   b. 2×10^6 pF
5. a. 3.7×10^7 μH
   b. 3.7×10^4 mH
7. $20°C$
9. 1.26 MJ
11. 8.3 msec
13. 5×10^6 m

UNIT 29 COMBINED PROBLEMS ON MEASURE

1. 304.8 mm
3. 8.045 km
5. 22.712 L
7. 11.468 m³
9. 20°
11. 36 MJ
13. 0.2 MV
15. 1 245.443 L
17. 7.079 m³
19. 0.4 cm
21. 0.75 cm
23. 25 m
25. 0.950 mm²

SECTION 7 RATIO AND PROPORTION

UNIT 30 RATIO

1. a. 3:1
   b. 5:2
   c. 3:1
   d. 3:1
   e. 1:3
3. 100:9
5. 1:7
7. 7:36

UNIT 31 PROPORTION

1. 61.2 min
3. 11.52 min
5. 3.192Ω
7. 2.59Ω
9. $33.10
11. $218.02

UNIT 32 COMBINED OPERATIONS WITH RATIO AND PROPORTION

1. $75
3. $400
5. 1:2
7. 1:20
9. 900 r/min

SECTION 8 FORMULAS

UNIT 33 REPRESENTATION IN FORMULAS

1. $R_t = R_1 + R_2 + \ldots + R_n$
3. $\dfrac{E_p}{E_s} = \dfrac{P_p}{P_s}$
5. $C = \dfrac{1}{2\pi f X_c}$
7. $P = \dfrac{E^2}{R}$
9. $M = k\sqrt{L_1 L_2}$

11. Conductance (G) is equal to the reciprocal of resistance.
13. The effective value of an AC voltage (E) is equal to the maximum peak value (E_{max}) multiplied by 0.707.
15. The frequency (f) of an AC generator can be calculated by dividing the product of the number of pairs of poles (P) and the speed of the generator in revolutions per minute (N) by 60.

UNIT 34 REARRANGEMENT IN FORMULAS

1. $C = \dfrac{Q}{V}$
3. $Z = \sqrt{R^2 + X^2}$
5. $R_t = R_1 + R_2 + R_3$
7. $C = \dfrac{1}{2\pi f X_c}$
9. $N_p = \dfrac{N_s E_p}{E_s}$
11. $Z_p = \dfrac{Z_s N_p^2}{N_s}$
13. $r_p = \dfrac{u}{g_m}$
15. $k = \dfrac{M}{\sqrt{L_1 L_2}}$

UNIT 35 GENERAL SIMPLE FORMULAS

1. 40.68 Ω
3. 28.52 kW
5. 1.57 A
7. 12.5 A
9. 392°F
11. 149°F
13. 50 Ω
15. 282.5 r/min
17. 6 171 W
19. 0.926 A
21. 46.77 A
23. 12.37 Ω
25. 80 Ω
27. 83.83 kW
29. 29.92 kW
31. 91.77 A
33. 21.19 A
35. 73 252.17 CM
37. 60 Hz

UNIT 36 OHM'S LAW FORMULA

1. 0.52 A
3. 16.70 A
5. 11 Ω
7. 1.37 A
9. 0.52 A
11. 240.01 ft
13. 2 637.2 ft
15. 11.2 Ω
17. 0.6 Ω
19. 7.33 Ω
21. 0.25 Ω
23. 0.936 Ω
25. 101 323.64 CM

UNIT 37 POWER FORMULA

1. 3 450 W
3. 0.652 A
5. 1 650 W
7. 114.94 V
9. 113.64 V
11. 3.48 A
13. 28.8 Ω
15. 2 332.27 W

UNIT 38 COMBINED PROBLEMS ON FORMULAS

1. 2.1 A
3. 20 Ω
5. 0.96 W
7. 31.5 ft
9. 0.548 A
11. 4.3 A
13. 633.6 W
15. 21.193 A
17. 4 400 Ω

SECTION 9 TRIGONOMETRY

UNIT 39 PYTHAGOREAN THEOREM

1. 19.21
3. 51.96
5. 42.55
7. 59.24 ft
9. 12.17 mi

UNIT 40 TRIGONOMETRIC FUNCTIONS

1. 45.6°
3. 31.3°

5. 16.8 m
7. 1.3 m

9. a. 20.6°
   b. 159.4°

UNIT 41 PLANE VECTORS

1. 36.06 Ω
3. 17.32 Ω
5. 22.80 V

7. 250 V
9. 127.37 V

11. 120 V
13. 5 A

UNIT 42 ROTATING VECTORS

1. 8.68 A
3. 160 V
5. 86.58 A

7. a. 0 V
   b. 51.76 V
   c. 173.2 V

7. d. 193.18 V
   e. 200 V
9. 100 kHz

UNIT 43 COMBINED PROBLEMS ON TRIGONOMETRY

1. 8.94
3. 287.23
5. 66.66

7. 77.9 Ω
9. 12.5 Ω
11. 207.8 V

13. 15.5 A
15. 120°

ACHIEVEMENT REVIEW A

1. 65
3. 4 473 W
5. 2 1/48 hp
7. 3.586 A
9. 1.341 in
11. 3.672 8 cm

13. 2.392 in
15. $739.40
17. 262.812 W
19. 81.774 W
21. 4.026 Ω
23. 10 A

25. 2.2×10^{-6} F
27. 100 Ω
29. 150 V
31. 628.32 Ω
33. 51 000 W

ACHIEVEMENT REVIEW B

1. $226
3. 6 wk
5. 7
7. 6 945 r
9. 3 A
11. 8.553 in
13. 0.28 in

15. $2 730
17. 50 μF
19. 16 Ω
21. 18 in
23. $r = \dfrac{EN - IR}{IN}$

25. 55.287 A
27. 4 m
29. 50 ft
31. 40 Ω
33. 5.4 m